ON THE NATURE OF RIVERS

ON THE NATURE OF RIVERS

with case stories of Nile, Zaire and Amazon

Essay by

Julian Rzóska

'L'acqua che tocchi dei fiumi e l'ultima di quella che
andò e la prima di quella che viene.
Cosi il tempo presente'.

In rivers the water you touch is the last of what has
passed and the first of that which comes.
So with time present.

Leonardo da Vinci, 1452–1519
Codex Trivulziano fol 34 r., Milan

Springer-Science+Business Media, B.V. 1978

On cover. The Nile in the desert, space photograph taken from an altitude of about 300 kilometers, before the Aswan High Dam. In the lower right corner the junction of the two Niles is discernible; the joint river with the great bend is visible, just short of the Luxor–Qena bend, the Red Sea, Yemen and Saudi Arabia stretch towards the curvature of the earth. Courtesy of NASA.

ISBN 978-90-481-8517-7   ISBN 978-94-017-2480-7 (eBook)
DOI 10.1007/978-94-017-2480-7

Cover design Max Velthuijs

To Jack F. Talling, F.R.S.
friend and companion in river research

Contents

Foreword

Origin and Aims of this essay

It was not by choice but by the misfortunes or fortunes of the last war, that I became involved with rivers. In December 1946 I obtained a lecturing post at the then Gordon Memorial College at Khartoum and the Principal of the college brought me to confluence of the two Niles and urged me to 'do something' on the biology of the river. I was very reluctant, my experience was limited to lakes in Poland up to 1939, and I did not know anything about work on rivers. The 'equipment' was a rowing boat, hired, and a 'home made' plankton net. This limited our first exploratory steps to the immediate vicinity of Khartoum. In both the White and Blue Nile we discovered the presence of a pure plankton. This was contrary to opinions expressed in the limited scientific literature available at Khartoum which stimulated our doubts and the search for the origin of this phenomenon. And so, early in our work, we became aware of the longitudinal sequence of events in running water, a fundamental feature of river ecology.

In the Nile, as in other long rivers, the difficulties of work were daunting; the water courses stretch for thousands of kilometers south and north of our base, our work had to be done in time free from lecturing duties, no research grants were available.

But experience and equipment grew, above all the number of collaborators. Tribute is due to them: A. Brook, the late G. Prowse, J. F. Talling, I. Thornton and P. Gay. Their knowledge of algal systematics, ecology and chemistry enlarged my limited horizon. In 1953 a launch was acquired, transformed from an old Shell-tanker, and in that year the 'Hydrobiological Research Unit of the University of Khartoum' began to publish its Annual Reports. The initial team dispersed and D. Hammerton took over and assembled new collaborators. It is gratifying to know that after his departure Sudanese biologists keep the work alive.

In 1958 I left the Sudan, some of our results were published, more were to come. The spell of the great river with its history never left; J. F. Talling deepened our knowledge of the river system decisively by his research on the African limnological scene. Neither did the spell leave me, though my new teaching duties in London were very heavy. Two issues were before my eyes: the incongruence of understanding the nature of rivers and the possibility of bringing all the results of work on the Nile into a survey. Finally, with the help of willing co-authors I brought together the monograph on 'The Nile Biology of an Ancient River', published in March 1976 by Junk at the Hague. While assembling and writing this book, I became even more conscious of the total inadequacy of the present representation of rivers in hydrobiology. This challenge is now responded to in these reflections.

1

I chose a simple title because the written language of science has become cumbersome; I call it an 'essay' because it is not anything more. It contains two parts one deals with the essential features of three great rivers, the other is of polemic character and tries to rectify some erroneous opinions. These, I discovered, were in contrast to sound research work carried out in the past.

A special remark must be made on the illustrations; most are taken from the Nile book, some few are new. I regard illustrations not as mere adornment, I use them as additional documents of visual nature. Especially the space photographs bring out the successive interaction between land and water, which is another basic feature of river ecology. No suitable illustrations were found for the case of the Zaire (Congo) river, but two superb air photos contain the essential characteristics of the Amazon.

Acknowledgement

I owe a great gratitude to J. F. Talling who helped with advice and criticism; H. Sioli assisted with bibliography and friendly encouragement; so did G. Marlier; H. O'Reilly-Sternberg lent me an excellent photograph.

As usually the archivist of the Linnean Society in London, G. Bridson, helped with books; Miss Dorabella Northcott drew maps and read the proofs, an important task which I could not do because of an immobilised right hand.

Autumn, 1977
Julian Rzóska
6, Blakesley Avenue, London, W.5

Part I.

Essential features of rivers

Rivers in nature

Rivers drain parts of the earth's crust and in many cases are old features of the land. Geological and geomorphological events have shaped their courses and their present state can only be understood by their past. Similarly their characteristics are in close and mutual interdependence with the land they traverse. Climatic conditions govern their water supply via tributaries from long distances. The water mass in rivers is in motion, unidirectional and sometimes of great force. Their outstanding function is the transport of water; with changes of discharge and velocity of water movement they erode the land and carry sediment and often bedload for, sometimes, long distances. These sediments and bedloads are then deposited in downstream areas or are flown into the sea.

Rivers differ from standing waters not only in their continuous motion, but also in their longitudinal diversity; turbulent rocky stretches may alternate with quiet parts. By scouring the riverbed they acquire mineral components, solids or solutions, from a wide range of their basin. As the shore line of rivers is relatively much greater than in lakes, the intimate connection between subsequent landscapes and river water is enhanced. Above all the medium, river water, is continuously passing away and this creates profoundly different conditions for life.

Piercing the land, rivers are communication channels for organisms, which migrate and colonise. This applies also to man. He had to live near water and rivers gave more opportunity to travel and search. From the Palaeolithic onwards man gathered on the Nile; these loose gatherings evolved into a long sequence of civilisations. Modern man has interfered with rivers using them as waste disposal channels and the force of running water has been concentrated by dams. In fact few large rivers are still free flowing, natural phenomena.

To study and understand rivers, interdisciplinary cooperation is necessary and this is difficult to achieve. Studies on running waters are usually patchwork, concentrated on short stretches; only four studies have so far treated the entire flow of long rivers, the Rhine, Volga, Danube and the Nile. All of these have been altered by the interference of man. It is hoped that the enormous detailed work on the Amazon will produce a monograph.

Case stories, Nile, Zaire and Amazon

Two maps allow a general orientation into the three river systems. The hydrographic map of Africa omits many details but shows the contrast between the Nile and the Zaire. (Figs. 1 & 2.)

Of all rivers in Africa only the Nile flows north, thrusts through the dry Sahel and through the desert; some small streams from the Atlas moun-

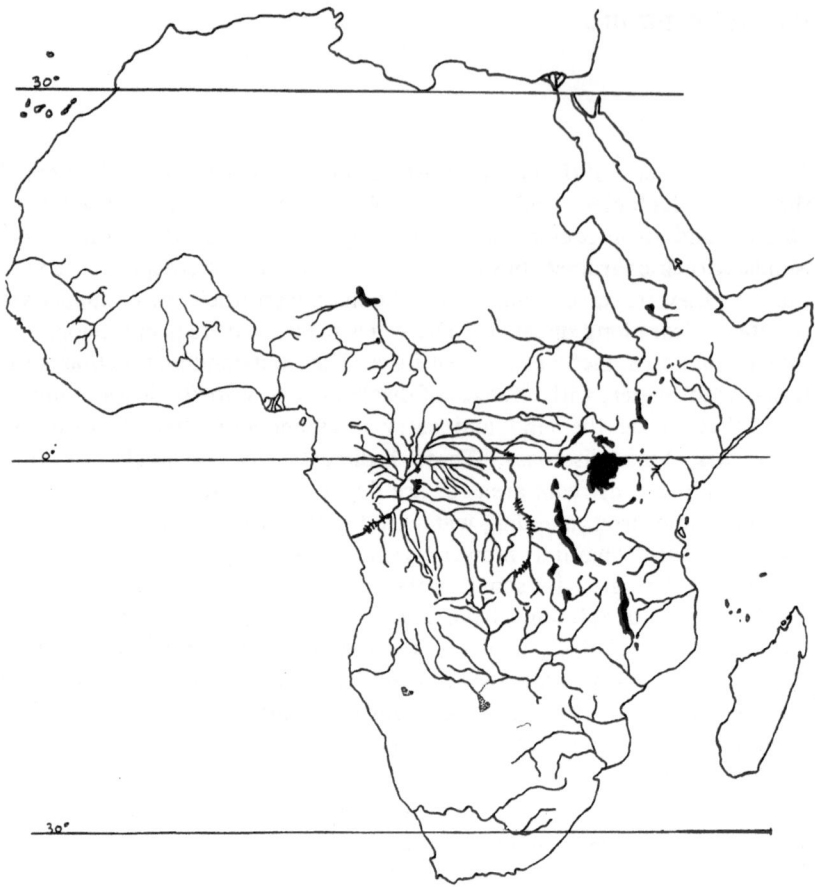

Fig. 1. Africa with the main river systems. The contrast is remarkable between the dense tributary net all along the course of the Zaire and its upper extension, the Lualaba, and the slender course of The Nile with its far away water supply. Note the difference of climatic zones traversed by Zaire and Nile, the latter thrusting north through the desert to the Mediterranean. Note also the isolation of the major lakes in contrast to the long courses of rivers. From various sources.

tains reach the Mediterranean. This extraordinary, freakish, behaviour can only be explained by the origin of the Nile from fragments. It has aptly been called a 'hydrological phenomenon', its importance for transporting water to the dry north is outstanding in the whole world; its hydrological regime is known in details, its water budget carefully calculated. The Nile is now a river managed by man for conservation of water, at present nothing is discharged in the sea.

The Zaire is a natural phenomenon with abundant equatorial water supply; it occupies about a quarter of the continent and is a dominating feature of Africa. Unlike the Nile, the Zaire is not managed by man, only a

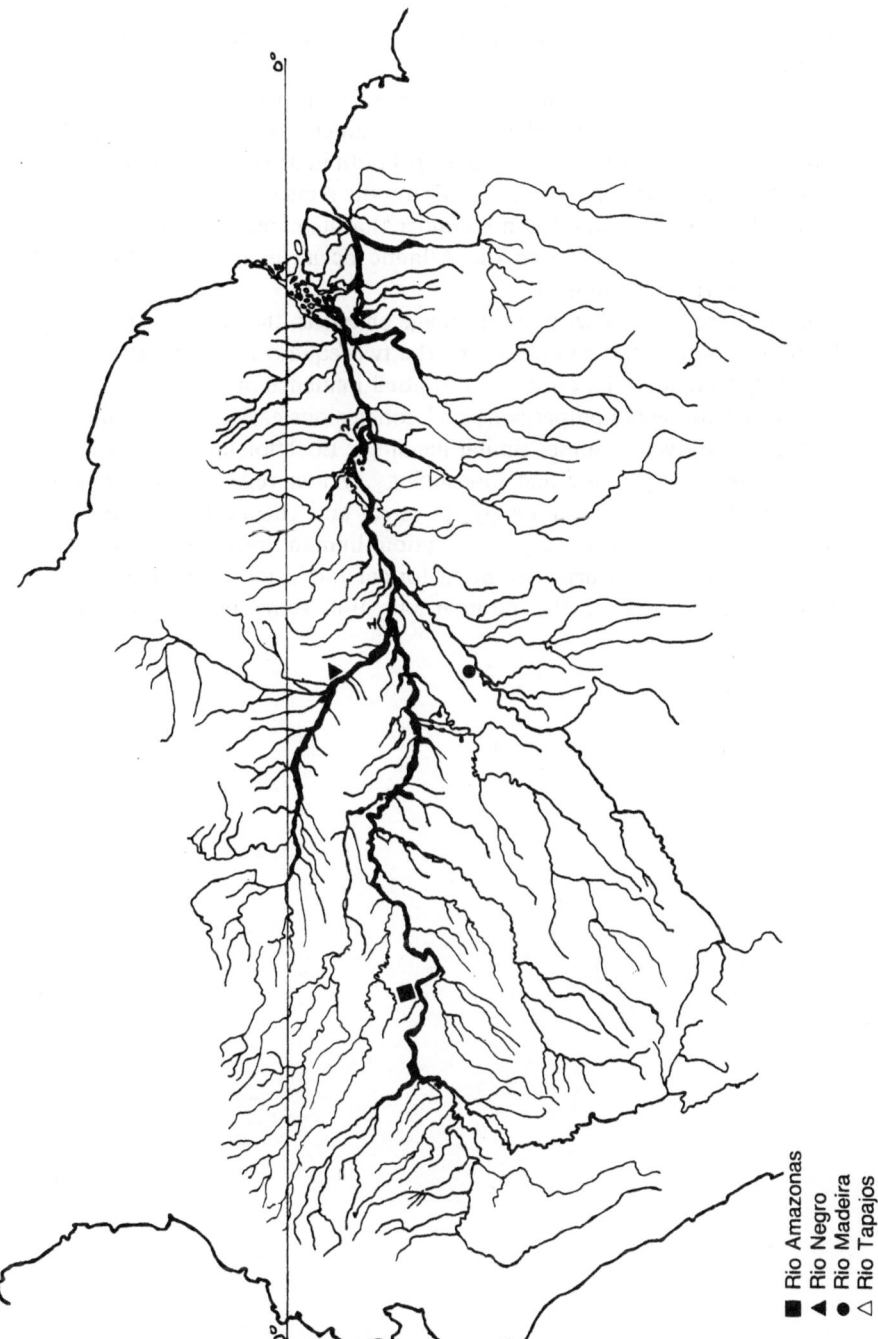

Fig. 2. The Amazon system dominating the South American continent. Its river valley, Amazonia, is bigger than the whole of western Europe. Only the biggest affluents are shown, 20 of them over 1,000 kilometers long. An equatorial tropical climate provides the enormous water discharge 4 times larger than the Zaire, more than 8 times that of the Mississippi and about 40 times larger than the discharge of the Nile. From various sources.

few small dams have been built for electric power in the industrialised south-east.

The extreme of river phenomena is reached in the Amazon, which imprints its dominance on a third of the continent of South America. The Amazon is the greatest river of the world both in its tributary network and its water mass. Even more than the Zaire, the Amazon has created its own series of landscapes and it is no wonder that its enormous basin is called Amazonia. Its discharge into the Atlantic ocean is as big as that of all other great rivers combined.

Both Zaire and Amazon are drainage channels, the Nile is a life artery. In comparison to the complexity of the two equatorial rivers, the Nile is relatively 'simple' in its course and mutual influence of land and water.

In order to make comparisons and conclusions easier, the treatment of all three case rivers is under similar headings: Location and area. Origin of river system, its geology and morphology. Water regime. Climate, plant cover and its impact. Limnology, physical and chemical characteristics. Biology of waters. Discovery, exploration, human presence and pressure.

The Nile has an unsurpassed prehistorical and historical documentation, more fully treated in the monograph. But some details had to be mentioned.

Case 1. The Nile

The map of Africa (Fig. 1) shows the extraordinary course of the Nile from the equator and Ethiopia to the Mediterranean, different from any other river in that continent. The river traverses 32 degrees of latitude but spreads only to 8 degrees of longitude. The composition of the river system is simple compared with the Zaire and Amazon; except the Atbara almost midway, all its tributary sources are far south. The length of the main axis from lake Victoria to the sea is 5,611 kilometers, not counting the Kagera river, the chief supply of lake Victoria. The theoretical drainage area is about 3 millions of sq. kilometers, of which 44% contribute nothing.

Origin, geology and morphology

Present geological evidence points to a late fusion of separate rivers in the final phase of the Pleistocene. The earliest traces of a river valley appear in Egypt in the middle Miocene; changing levels of the Mediterranean played a decisive role upon the slope and erosion of the river. Sediments of various origin together with crustal movements of rock formations allow to distinguish five phases of the Egyptian Nile. Old geological formations exist only in traces, most of the Sudan and Egypt is covered by a Tertiary Nubian sandstone and sands of desert origin. Tectonic tilting of the East African Plateau contributed to the creation of lake Victoria and its outflow; faulting on a large scale formed the Rift valley and affected the direction of the Nile. Zoogeographic affinities of fishes and mollusks indicate former eastern connections of the Victoria Nile with lake Rudolf and other eastern valleys. The origin of the swamp basin, the Sudd, has elicited much speculation, but is believed to be only of late Pleistocene age. The White Nile north of the swamps is apparently an old drainage line. It is this region which caused an early observer, Lyons in 1909, to express the firm opinion that the Nile is a system of separate structures. The present Blue Nile, and the other two larger rivers, Sobat and Atbara, owe their direction to the rise of the volcanic High Plateau of Ethiopia probably in the Oligocene. Recent rock analyses have dated the outflows of lava into the Sudan plains at about 20 million years, and the Blue Nile gradually sheared its way downstream through the great gorge. There are strong indications from sediments that all these Ethiopian rivers flowed to the north independently from the upper parts of the river system and joined only late.

The morphology of the Nile basin is dictated largely by the slopes of the rivers. The White Nile descends from the East African Plateau at 1,134 meters to 457 meters of the Sudan plains over a distance of about 800 kilometers and then falls gradually to sea level over the rest of the main course. The Blue Nile has a simpler morphology; after leaving lake Tana at 1,829 meters it passes over a spectacular waterfall, plunges into a gorge

9

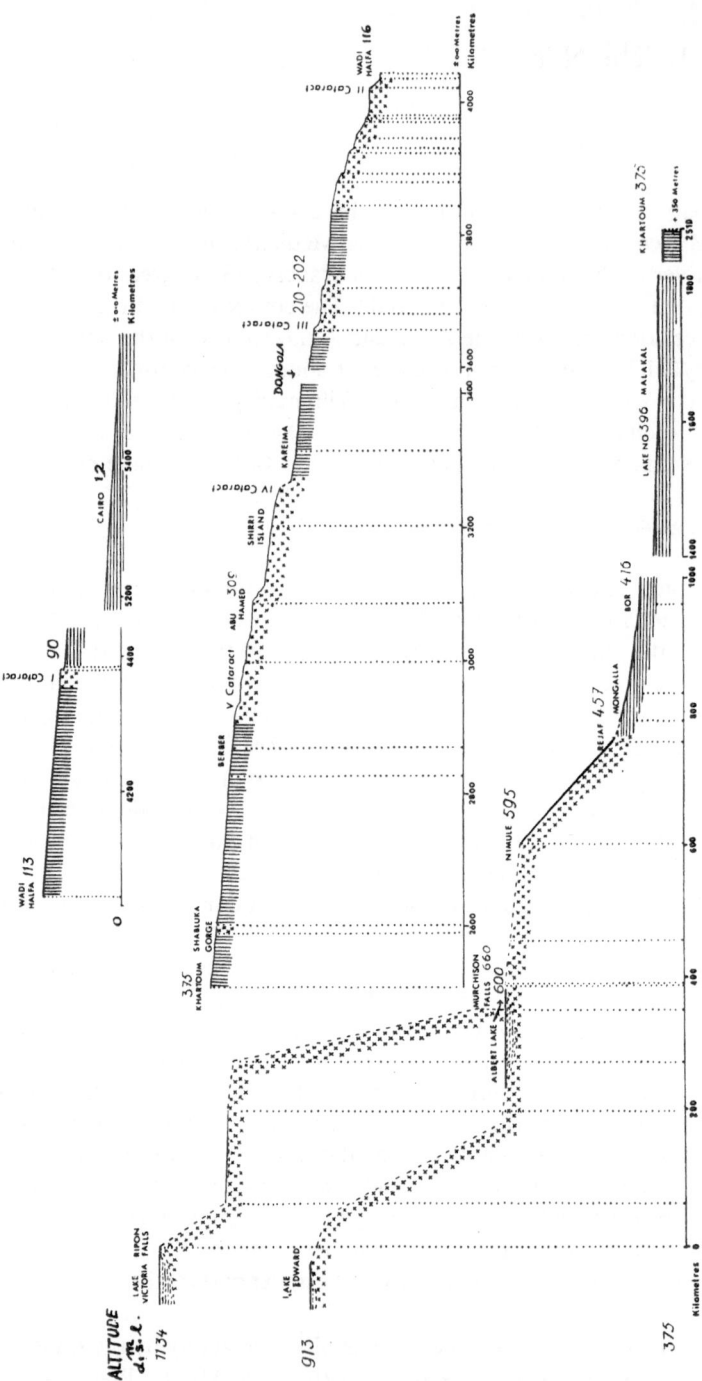

Fig. 3. Longitudinal profile of the main axis of the Nile from lake Victoria to the Mediterranean. Note the descent from the East African Plateau and the rocky outcrops of the cataracts. This profile induced Lyons (1909) to propose a multiple origin of the Nile from fragments. Adapted from Lyons.

10

800 kilometers long and in parts 2,000 meters deep, enters the Sudan plains at about 500 meters through short rocky stretches and flows further 800 kilometers through its own alluvial sediments to the junction at Khartoum at 372 meters altitude. (Figs. 3 & 4.)

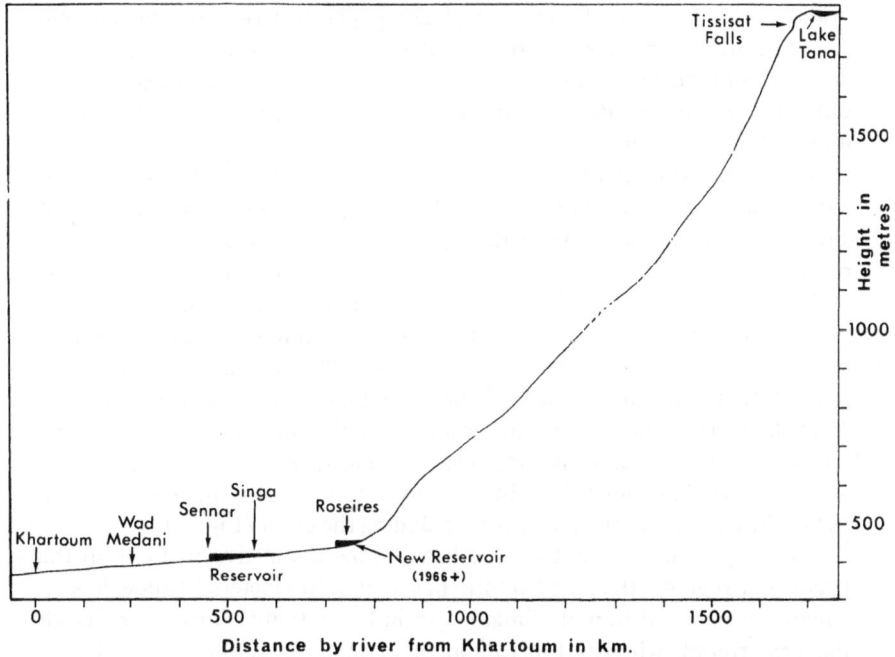

Fig. 4. Profile of the Blue Nile from lake Tana to the confluence at Khartoum. Note the steep descents through the gorge and the position of the reservoirs. Adapted from Talling & Rzóska 1967.

The joint river flows for 3,000 kilometers through a series of 'cataracts' or rocky sills; the most spectacular are the geologically puzzling Sabaloka gorge, the II cataract, now drowned and the Aswan cataract now the basis of the great dam. These cataracts interrupt the flow of the river abruptly. The river valley is surrounded in the last 3,000 kilometers by desert, leaving only a narrow fringe of riverain life giving soil. Tracts of alluvial soils are spread across some parts of the central Sudan as the result of Blue Nile inundations. In general the soils of Africa are poor.

Climate past and present

In sharp contrast to the two other rivers discussed in further pages, the Nile is and has been dominated by climatic changes. These must have existed in the geological past as attested by gravel deposition and their location under more recent sediments; their interpretation belongs to geologists. More information is available for the last 30,000 years, in which shifts of rainfall occurred. Butzer (1966) recognises a series of wet

11

and dry phases in northern Africa affecting Egypt and the present Sahara. Wickens (1975) has traced recently the 'changes in the climate and vegetation of the Sudan since 20,000 B.P.' For us the most important part of these changes are those which lead to the present state of aridity over a great area of the Nile basin. A rich documentation for this last phase of aridity exists in rock drawings and camp places from the Palaeolithic onwards, intensified during pre- and dynastic times of ancient Egypt by pictorial and written records. We owe much of the evidence of the drying out of the surrounding lands and the advance of the desert to archaeologists and historians.

At present the great desert has enclosed 3,000 kilometers of the northern basin from both sides with radical effects on life in every form. Rainfall along the Nile diminishes from the equatorial lake Victoria with 1,500 mm per average year with two annual peaks, to 25 mm and nil over the last 1,500 kilometers of the Sudan and Egypt, which has only a narrow coastal belt of Mediterranean rain. The Ethiopian Plateau has a different climatic regime. The rain is of the monsoon type, with 90% of the c. 2,000 mm of precipitation concentrated from June to October with a peak in July and August. A large part of the rain drains through numerous tributaries into the Blue Nile gorge causing the great sediment laden flood. The sediments are transported all along the joint river to Egypt and they have created the fertile alluvial soils of the Nile valley including the delta. (Fig. 5.)

Air temperatures over the Nile basin show distinctive features in the headwater regions; the lake Victoria basin, at 1,300 meters altitude, has an equably warm and humid climate with little fluctuation around 25°C all the year round, whereas the Ethiopian Plateau, at 1,800 meters at lake Tana, fluctuates between extremes of 23 to 30°C during the day to 6 to 8°C at night. With the descent into the Sudan plains both rivers and their joint course are subjected to increasing high temperatures and increasing seasonal and nocturnal variations. Extremes are reached in the desert climate of Nubia and upper Egypt where the thermometer may reach over 40°C and winter nights may fall occasionally to freezing point. Relative humidities vary between 80% over the high altitude head regions to 20% and even lower over the desert. This climatic regime causes great changes in the longitudinal sequence of life in the three climatic zones which the river system traverses – a decisive contrast to the other two rivers discussed.

Water regime

The hydrology of the Nile is also quite different from the Zaire and Amazon. On the whole the river is poor in water resources over the year, but seasonally fluctuates between scarcity and abundance. A map shows the origin of water in the river system from three main regions. Ethiopia, the lake Victoria basin and from the Nile–Zaire divide. The latter supply is minimal; of all the small rivers issuing there only the Ghazal reaches the main river, the others peter out in swamps. The Ethiopian contribution is

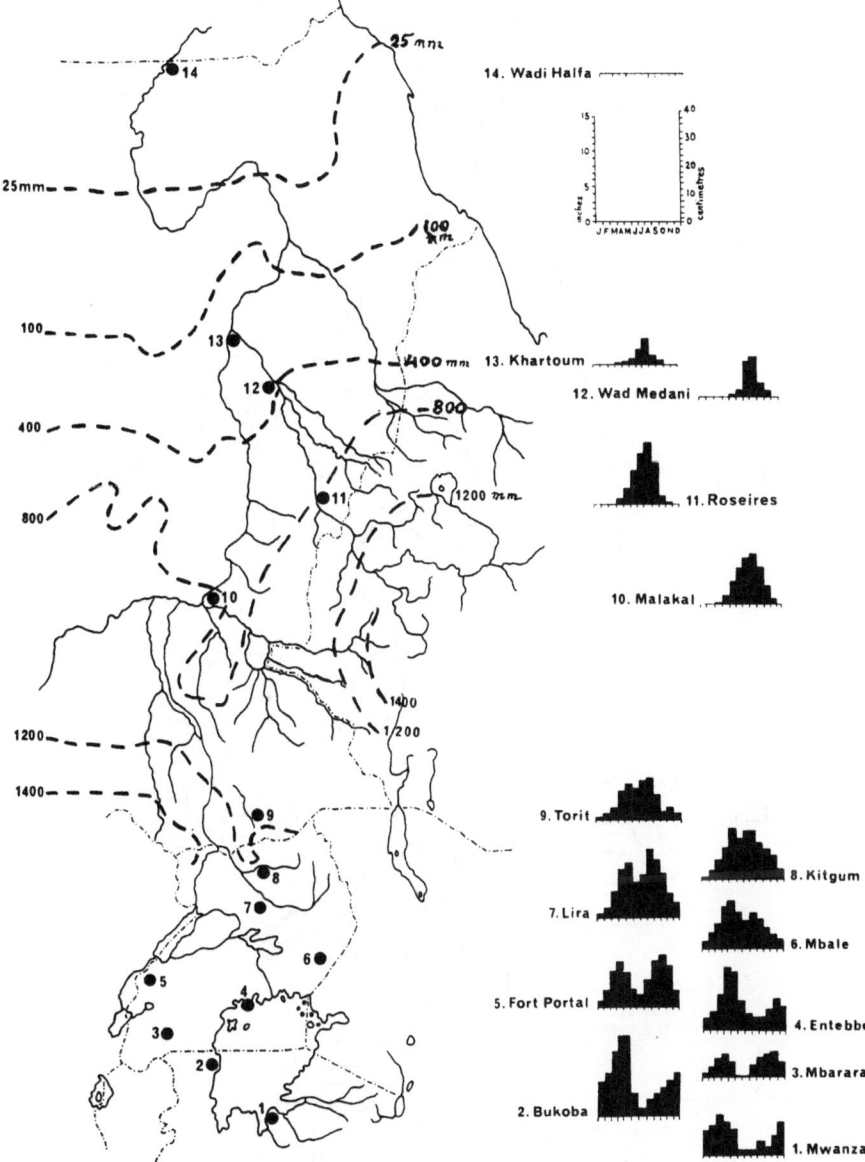

Fig. 5. Rainfall over the largest part of the Nile basin. It dwindles from 1,500 mm over lake Victoria and about 1,200 mm over lake Tana to about 160 mm at Khartoum and 25 mm over the great bend of the Nile; further north no regular rainfall exists. Note the gradual constriction of rain into a limited season of few weeks. This dominates life in the Nile basin. From various sources.

the biggest with the Blue Nile and its tributaries the Dinder and Rahad, and the Atbara. Except the Ghazal supply, all others come from countries outside the Sudan; Egypt has only the Nile for its water. (Fig. 6.)

13

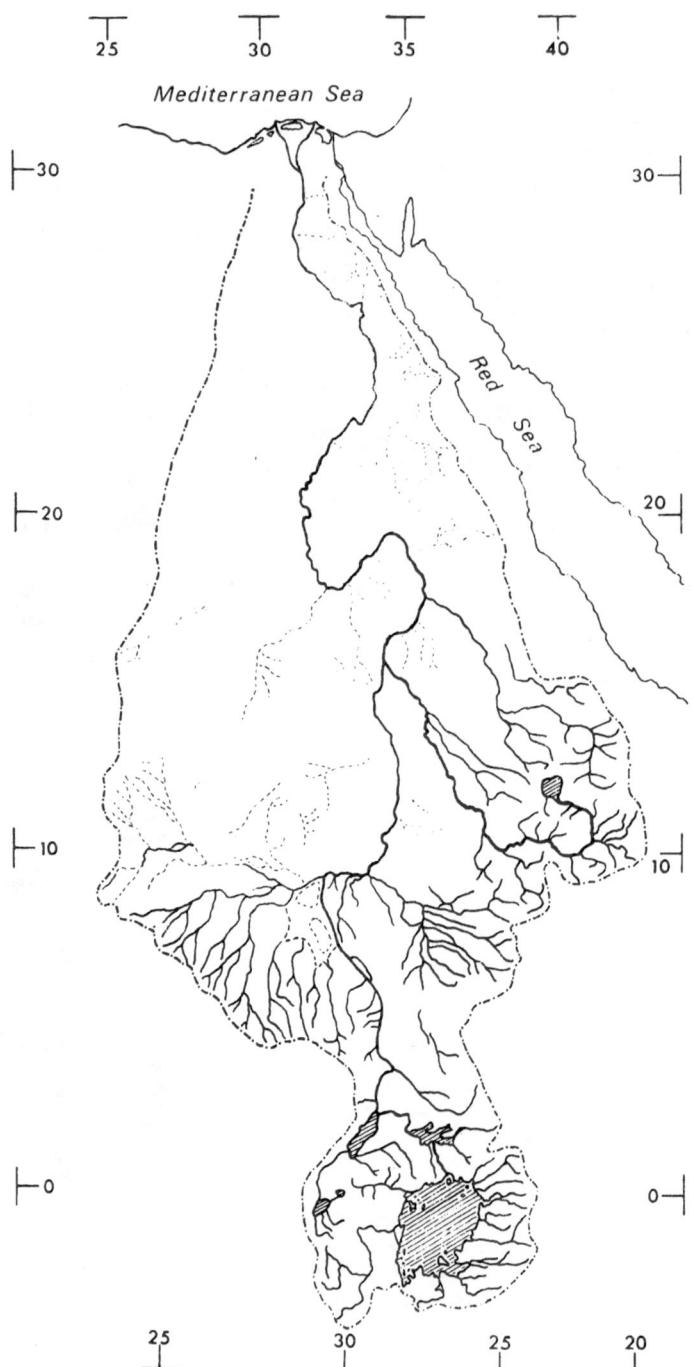

Fig. 6. Hydrographic map of the Nile system. The total drainage area is indicated by a dashed line (-·-); 44% of the area contribute nothing; all areas of water supply lie outside the two consumer countries, Sudan and Egypt. Note the ancient water courses now extinct indicated by . . . . . lines. Adapted from Lockermann (unpublished thesis).

A graph gives the contribution of the three main components to the total water regime. It shows the seasonal scarcity from December to June and the peak of supply from July to October. (Fig. 7.)

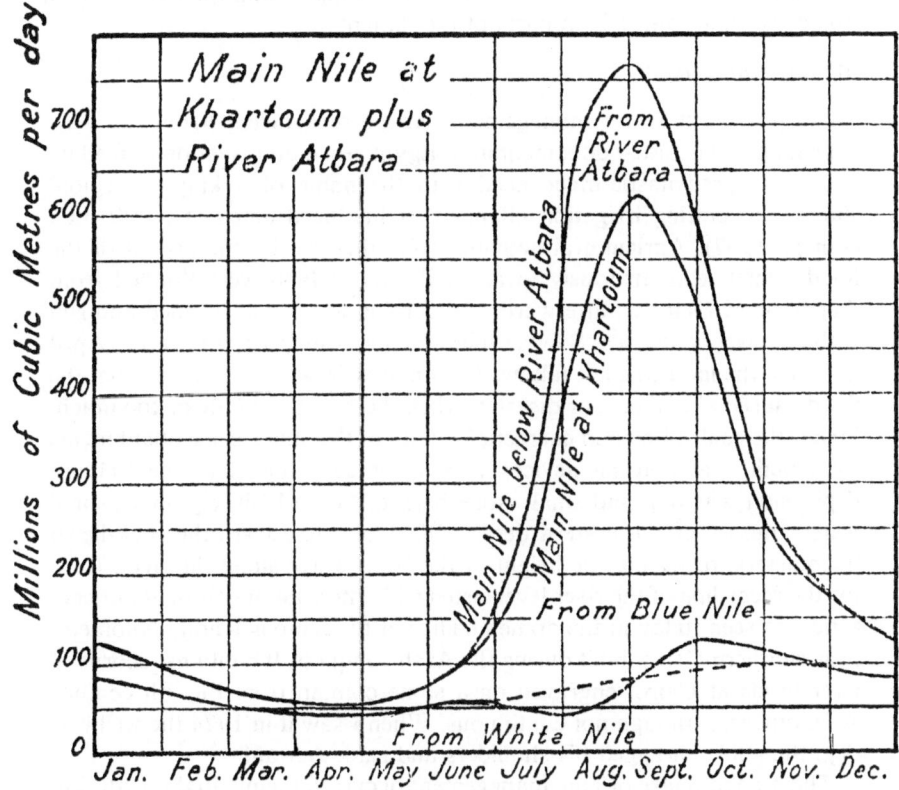

Fig. 7. Graph showing the water contribution of the main components of the Nile based on observations from 1912 to 1936. Although wide fluctuations of discharges exist, the long term averages of discharge are valid at present. From Hurst 1952.

The graph gives the monthly discharges over a period of 24 years and does not include the effect of the big dams. It shows above all the unevenness of the supply, too much in July to October, too litttle in the other months. It attests further to the role of the Ethiopian supply from the floods of the Blue Nile and the Atbara river and the much lesser role of the White Nile.

The total annual average is, according to official Sudanese statistics, in 10^9 m^3 (thousands of millions of cubic meters):

From White Nile		From Blue Nile
23		51
Main river at Khartoum	74	
Atbara	12	
Total	86	(85 according to Egyptian sources)

15

These are averages for 60 years; but averages do not reflect the fluctuations of discharges in exceptional years, with usually grave consequences. It should be stressed once more that both Egypt and the Sudan are agricultural countries and rely on water supply from the river, though some seasonal rain falls in parts of the Sudan.

Water management

Three phases of water management can be recognised. From the very beginning of historical documentation Egypt relied on irrigation to feed its people. A predynastic-mace head with the name of a king 'Scorpion' shows him cutting irrigation channels; it has been dated at 3,000 B.C. (Emery 1967). Agricultural activities were dictated by the arrival of the flood, which was anxiously awaited. Basins of land were flooded from August until October; afterwards the surplus water was drained and run with the river to the sea. Sowing followed, harvesting was in March–April and then the land was left fallow. The amount of water available from the flood caused abundance or disaster, when there was too little or too much. Under the centralised and energetic regimes of the pharaohs, embankments were built, water lifting devices were invented, a canal to the Faiyum depression was dug and small stone barrages were built, e.g. in seasonal water courses of the eastern desert. Of great interest are the rock-hewn 'Nilometers' recording the height of the flood water along the river. Care for the recording of the river level lasted with interruptions throughout centuries, as seen today in the domed edifice of the famous Rhoda Nilometer near southern Cairo, built during the Arab conquest. It contains records of river levels at Cairo, engraved on a stone column from the 7th century A.D. onwards, though not continuous. When I saw it in 1974 the well was dry; the guide shrugged his shoulders and said: Aswan!

The second phase of Nile management occurred mainly in the 19th century. Napoleon Bonaparte's engineers and scientists, who accompanied his invading army in 1789, measured the deposition of sediments (accurately), devised better distribution schemes of water and engineering works, besides recording for the first time the antiquities. Political turmoil prevented the practical application of their ideas, yet the 24 volumes of the 'Description d'Egypte' published in Paris in 1820 remain a milestone in the new era. From 1845 the pace of Nile water regulation increased, barrages were built at a number of places on the river. These stored surplus water to help out in the lean months. In 1902 the first dam of solid masonry was erected across the first cataract at Aswan, to store more water and for a longer time for the increased needs of a growing population.

The 20th century saw the onset of the last and present phase of Nile management. The prophetic words of a gifted journalist and later politician written in 1903, on the occasion of the first Aswan dam, heralded the new era: 'of even mightier schemes, until at last every drop of water, which drains into the whole valley of the Nile shall be equally and amicably divided ... and the Nile itself ... shall perish gloriously ... and never reach the sea. . .'.

16

Sixty years later the prophecy of Winston Churchill took practical fulfillment. But in 1904 Willcox estimated that only a third of the Nile water could be used, two thirds had to be discharged into the sea. He described the breaching of river banks by a disastrous flood in 1887. A grave dilemma faced the Egyptian government, a rapidly rising population, agriculture expanding by necessity and the water necessary not available at the right time. Then there was the southern neighbour, the Sudan, emerging as consumer of increasing quantities of water. The Department of Public Works in Egypt, under a group of able hydrologists and engineers began its work, collecting data on discharges over the whole river system incorporated in the truly monumental volumes of 'The Nile Basin'. Blueprints for a rational administration of the resources were drawn up. In the last decades the Sudan government set up its own irrigation department and with some political frictions resolved, the two countries started to share the carefully measured Nile water. Dam projects were gradually realised.

Five dams operate now along the Nile, a sixth has been built on the Atbara river. Discounting the early barrages and the old Aswan dam, these are: the Sennar dam (1925) on the Blue Nile, the Gebel Aulia dam (1936) on the White Nile which serves Egypt, the Owen Falls dam on the Victoria Nile (1954), the dam at Khasm el Girba on the Atbara river (1964), the Roseires dam at the entry of the Blue Nile into the Sudan (1966). Finally the great Aswan High Dam replaced the old one, and began to fill in 1964; by now the dam is full. The Owen Falls dam is solely used for generating power, all others retain water for agricultural purposes with power stations at three of them. All of them have been constructed after preparations and much thought. This is especially true of the much maligned Aswan High Dam, which was contemplated already in the thirties of this century.. It is the only long term retention reservoir; the others are seasonal. The Aswan High Dam stores and regulates the wildly fluctuating floods and provides Egypt with a steady supply all the year round. In a recent symposium on 'Nile water and lake-dam projects' held in Cairo in March 1976, many problems of the river were discussed. A Swedish consultant presented a new balance sheet of available water resources and their use.

Table 1.
Balance Sheet of water resources of the Nile, in 10^9 m^3

	Without new dam	With new dam
Total water available	85.0	85.0
Allocation to Sudan	−13.0	−18.5
Evaporation losses in new dam	—	−11.0
Water passing Aswan	72.0	55.5
Irrigation Upper Egypt	−18.8	−24.3
Recirculation gains	8.0	5.3
Water passing Cairo	61.2	36.5
Irrigation in Delta	−28.2	−36.5
Discharge into Mediterranean	−33.0	nil
Total balance	0	0

Fig. 8. The southern end of the Blue Nile gorge, now a broad valley, surrounded by mountains. The entry of the powerful Didessa river is in the middle background. Photograph by P. Morris.

18

These are, in simplified form, the facts about the water of the Nile at present. In a situation where the population in both main consumer countries have risen to 50 millions, and the allocation per head of people in Egypt has actually fallen, obviously the utmost economy is necessary. I attended the Cairo meeting and I was impressed by the frankness with which the drawbacks of the enormous dam basin have been discussed. These are numerous but remedial measures are proposed and part already applied. These have been discussed by me recently (Rzóska 1976).

Inspite of the present dams, the cutting of discharge into the sea and water saving practices, the shortage of water continues. A new scheme, but contemplated for a long time, is under way. This is a canal by-passing most of the Upper Nile swamps, where half of the incoming water is 'lost' by spillage and evaporation. The scheme is controversial as it will change the character of a great flood plain, used by half a million people of cattle-owning tribes for grazing.

This chapter is to prove the almost desperate need to provide more water from this river of historic destiny. The Nile contrasts fundamentally to the Zaire and Amazon.

Vegetation and landscapes in the Nile valley

Altitudes, rainfall, span of temperatures and water regime show drastic changes from the equator to the Mediterranean, and vegetation follows suit. The East African lake plateau is now a mixture of savanna-woodland and only remnants of the primeval forest exist. Ethiopia was covered partly by montane vegetation, which persists in the high mountains but is altered by man around lake Tana; the Blue Nile gorge is a unique habitat of its own, has a specific climatic regime, but above all is subject to the full onslaught of the torrential rush of water during the flood, followed by aridity. A fringe of forest and scrub extends along the narrow parts but widens downstream. (Fig. 8.)

In the Sudan belts of lowland forest and pockets of montane vegetation are followed north by savanna-woodlands, swamp and wetland savanna, then thorn savanna and finally semi-desert and desert. This reflects clearly the diminishing rainfall. (Fig. 9.)

The desert encloses the alluvial cultivated valley of the Nile in Nubia and Egypt, as revealed in the space photograph of the Nile bend at Luxor. (Fig. 10.)

The narrow river valley broadens into the great riverain oasis of Faiyum and finally into the delta, both intensively cultivated for thousands of years. (Fig. 11.)

A thin belt of coastal natural vegetation stretches along the Mediterranean.

Through the interaction of the factors mentioned, a series of landscapes surrounds the river system. Lakes from the sources of the two main component rivers; waterfalls, canyons and turbulent rocky slopes are followed by quiet, flat river stretches with inundations and swamps. These quiet

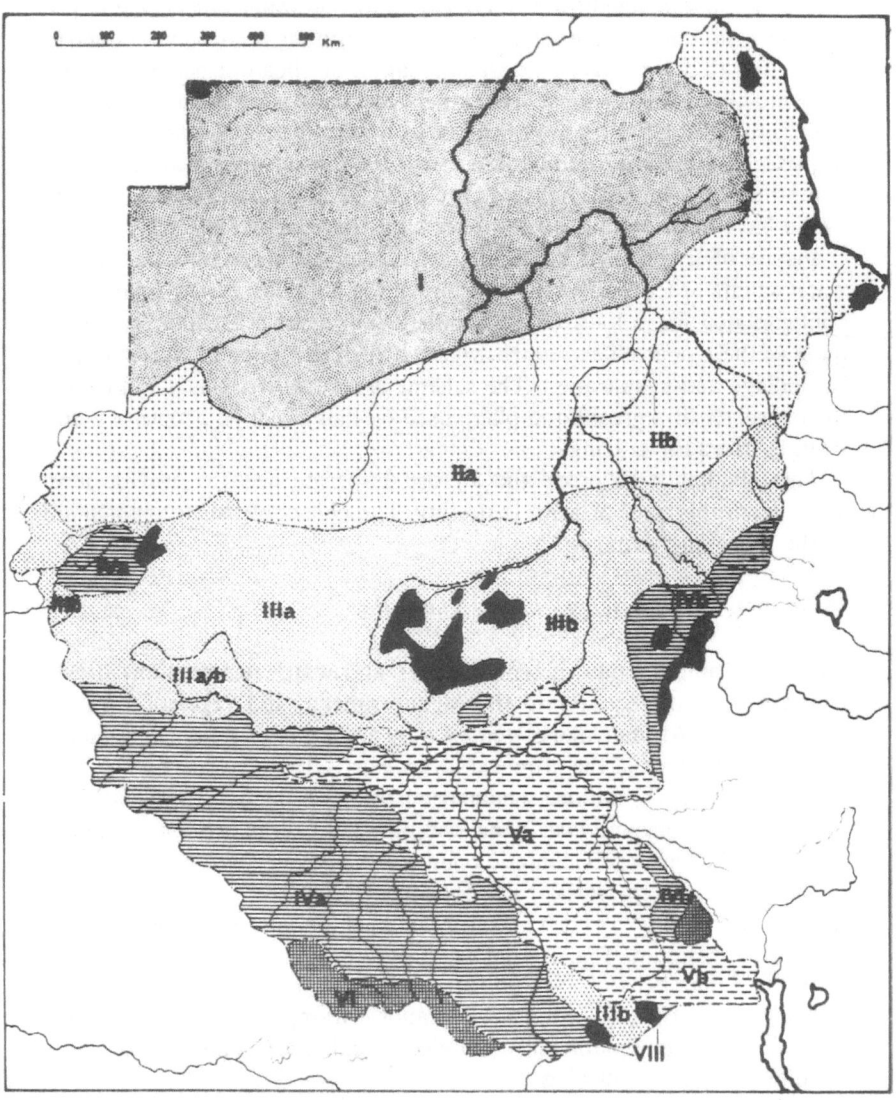

Fig. 9. Vegetation map of the Sudan enclosing the Nile. Zone I and II are desert or semidesert and occupy half of the country. Zone III is thorn savanna; IV represents deciduous savanna-woodland; V is swamps and wetlands, VI, VII and VIII lowland forest and montane vegetation (in small enclaves). From Wickens 1975.

parts of the river are interrupted again by rocky sills causing cataract conditions and gradually bare desert grips the river valley. Cultivation creates green areas where soils and irrigation allows and finds its climax in the Faiyum and the delta of Egypt.

Some of the more important landscapes of the Nile valley are illustrated in the pages of this essay. Lake Tana and the first part of the gorge of the Blue Nile, surrounded by mountains up to 3,000 meters high, is presented

20

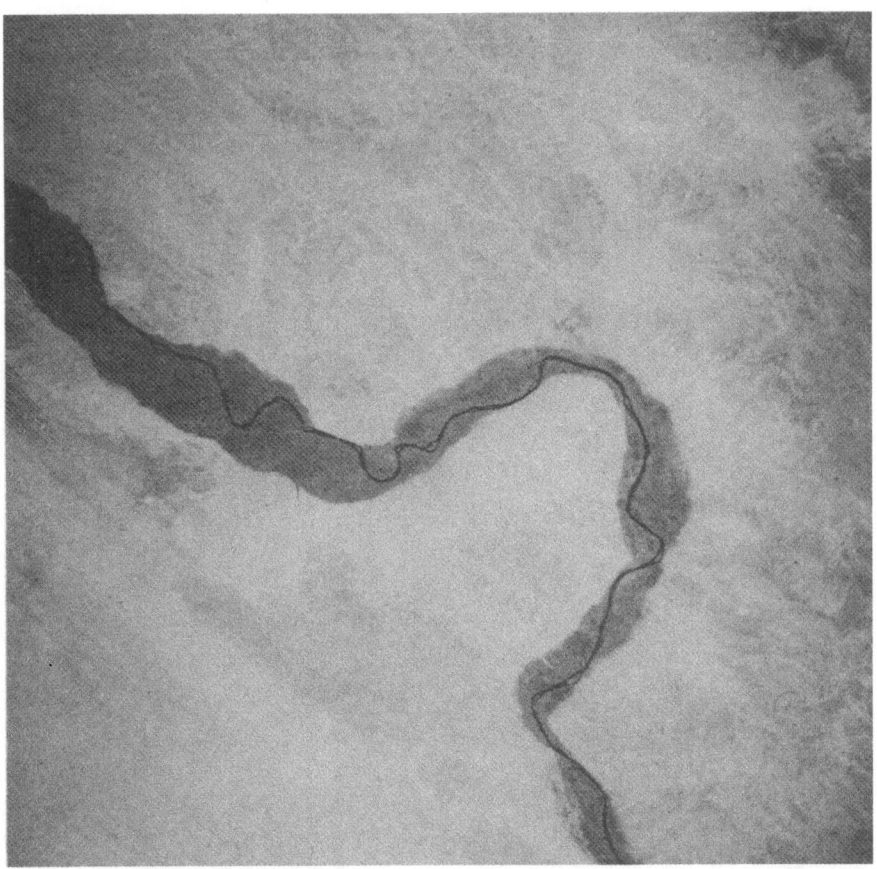

Fig. 10. The Nile valley from Luxor to Qena from space. Note the dark alluvial soil with the river in contrast to the desert. The river winds through the soils of its creation. Luxor with its famous antiquities is at the southern angle of the bend. Space photo ERTS.

in a space photograph. It clearly demonstrates the fundamental difference between a static water and a river forging its way through mountain slopes. (Fig. 12.)

This difference is reinforced by the two pictures (13 & 14) of the Tisisat Falls, about 30 kilometers below the lake Tana outlet. Here is water in violent motion, which creates both a landscape and a habitat. The abundance of water during the rainy season and the scarcity during the dry months is impressively seen. The end of the turbulent gorge was already shown in Fig. 8.

On the White Nile, the course of the river to the north of lake Victoria is seen in a space photograph. The Victoria Nile passes through lake Kioga, choked with papyrus swamps; its dendritic shape is a legacy from a drowned river valley, once a tributary from the east. (Fig. 15.)

21

Further north, at the escarpment of the East African plateau, the White Nile is shown forcing its way downstream through a narrow ravine into

Fig. 11. South-western part of the Egyptian delta, upstream of Cairo. The Fayium oasis cultivated for thousands of years, the brackish lake Qarun is shown. A streak of dark dots in the left middle border is the Wadi Natrun where in one of these depressions the last stand of papyrus was found a few years ago. Courtesy NASA.

the Sudan. We see here a sudden change from a placid, sluggish water course into a rapid, rocky stretch. (Fig. 16.)

One of the most distinctive environments exists in the Sudan plains, 800 kilometers downstream. The Upper Nile swamps represent a flat landscape of fringing papyrus, inlets, standing waters and vegetation-choked courses, quite different from any other environment on the Nile. These and the following illustrations demonstrate the succession of environments along the river system. (Figs. 17 & 18.)

22

The river and its valley as milieu for life

I have used the term 'life artery' for the Nile. This is mainly meant in the human sense; already the ancients recognised the exclusive role of the river

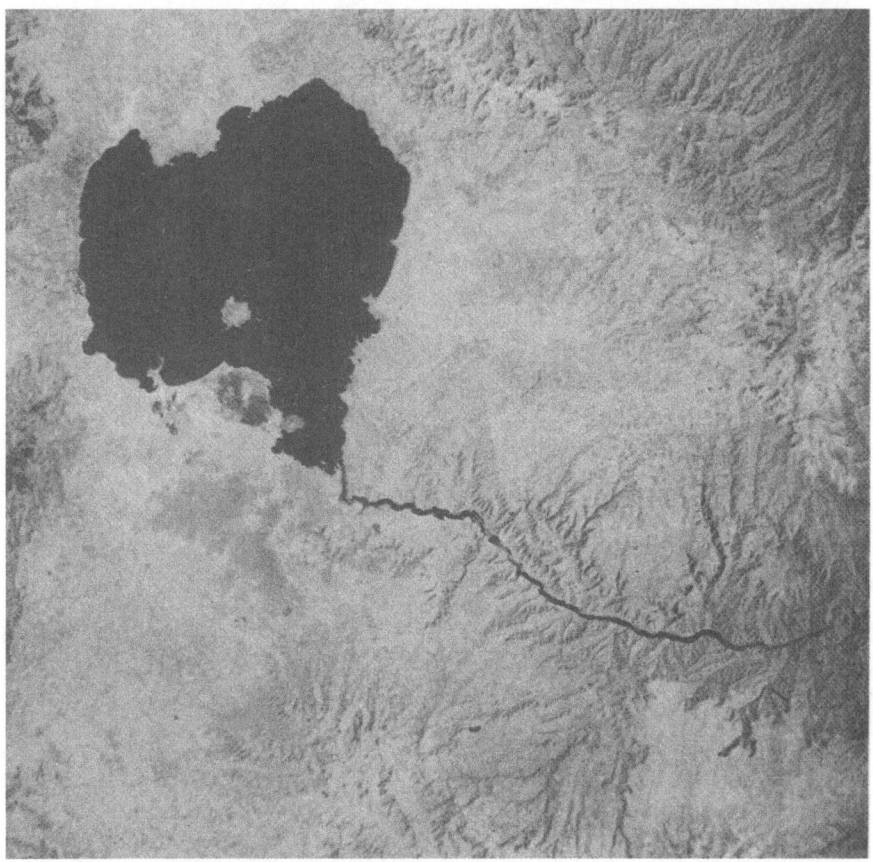

Fig. 12. Lake Tana in the Ethiopian High Plateau from space. The exit of the Blue Nile and the first part of the river gorge is shown. The numerous tributaries entering the gorge erode the valley and provide the great flood of the Nile and the sediment, which has built up the alluvial valley and the land of Egypt, 4,000 kilometers to the north. ERTS.

for the creation and sustanance of Egypt and its extraordinary civilisation. But for a much longer time the Nile linked Africa with the Mediterranean and served as a venue for plants and animals. In the land surrounding the river valley truly dramatic changes of climate have largely severed the African connection, yet enough remains for the link to be seen.

All over the north the great African fauna withdrew south before the onslaught of aridity, only bones and rock drawings of elephant,

23

Figs. 13 & 14. The Tisisat falls on the Blue Nile near lake Tana during the rainy and the dry season. The whole regime of the Blue Nile is governed by this enormous seasonal difference. Fig. 13 commercial; Fig 14 courtesy of Major Blashford-Snell.

rhinoceros, lion, giraffe, and even hippopotamus give evidence of former expanse. Only smaller species of mammals can live now in Egypt, mainly through human pressure. Palaearctic elements mix in Egypt with African species of birds, reptiles and amphibians. All these groups fan out towards the interior of Africa in a impressive way. The river itself harboured the

24

larger animals for a longer time; the hippopotamus was recorded in the delta by the great Buffon in 1769 (two specimens duly killed), and even as late as 1815 a single specimen was reported with incredulity by an Arab

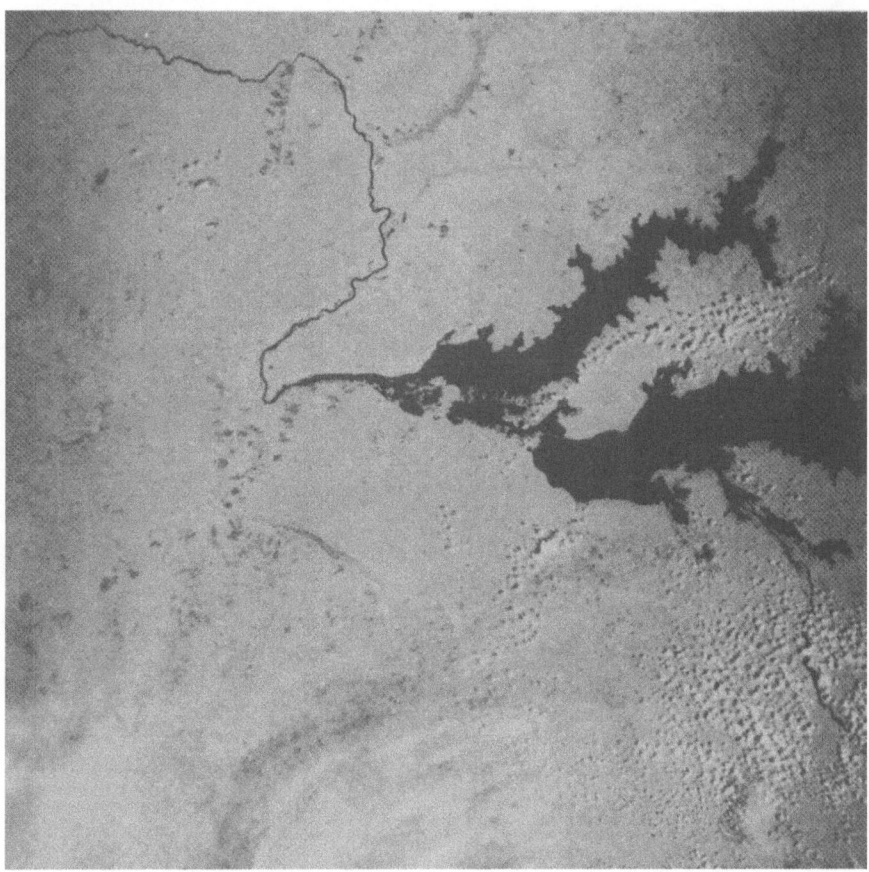

Fig. 15. Lake Kioga and the Victoria Nile from space. This a drowned river valley with a typical dendritic shape, a remnant of previous tributaries from eastern mountains. The lake is bordered by extensive swamps. ERTS.

chronicler. The crocodile, together with the hippopotamus, so frequently portrayed in the former papyrus swamps of ancient Egyptian murals, disappeared from lower Egypt but was still seen by travelers in its upper regions in the 19th century. The river milieu has preserved the connection with Africa longer and is still the most important link.

Fig. 16. The (White) Nile at Nimule nearing its sharp descent from the Ugandan escarpment into the Sudan plains. Up to this point it was a sluggish river bordered by swamps and extended for about 300 kilometers, before starting a new series of turbulent rocky stretches to the great swamps of the Sudanese Upper Nile. Photo, Sudan Survey.

Limnology and biology of Nile waters

Physical and chemical features

Foremost is the transport of water and its contents. No better testimony can be given to this than the space photograph of the delta. (Fig. 19.)

It looms dark against the surrounding deserts; apparently already the ancient Egyptians called their country the 'black land'. This is probably the most intensively cultivated land of the world. The papyrus swamps have given way to crops. This land was created by the sediments from

26

Fig. 17. Papyrus wall bordering lake No in the great swamps of the White Nile. Growths of *Vossia* and *Pharagmites* are interspersed, the vegetation wall reaches a height of 4 meters. Photo Rzóska.

Fig. 18. Inlets in the fringe of swamp vegetation are habitats of floating and submerged vegetation, here *Nymphaea* and *Trapa* are seen. These small bays harbour the richest community of plant and animal life in the whole river. Photo Talling.

Fig. 19. The Nile delta in Egypt from space. A wide panorama encompasses the Sinai pen-ninsula, the gulf of Eilat, the Red Sea, the Bitter lakes, the Suez canal and the canal to the delta. The four coastal lakes can be just discerned, also the Rosetta branch of the Nile. The difference of colour of the densely cultivated alluvial soil of the delta and the surrounding arid lands is staggering. Courtesy NASA.

Ethiopia through the carrying capacity of the Blue Nile and the Atbara rivers. See also Fig. 11 with the Faiyum.

Water temperatures follow the general climate and vary only slightly in the equatorial region of lake Victoria around 25°C; more differences have been observed in lake Tana. In the rivers the water mixes and stratification noted in the lakes is absent except in dam lakes. Seasonal differences increase north, but the whole amplitude is on average between 15 and 30°C, the lower values obtained rarely.

Stratification of water masses in the lakes and dam-lakes together with wind action causes overturns, affecting oxygen conditions and nutrient mixing; this was observed best in lake Victoria and also in lake Nasser, the northern part of the Aswan High Dam basin.

Transparencies in the Nile system vary greatly. The clearest water is found in lake Victoria with an euphotic zone of between 12 to 21 meters; the other high value for light penetration has been recorded in lake Nasser, up to 7 meters. In the rivers they are obviously much lower and rarely reach values higher than one meter; during the flood they may fall to almost nil.

Numerous data on conductivity show the local influence of waters from various regions; lake Victoria water has a low mineral content of about 65 mg/l whereas lake Albert has 480 mg/l due to the influence of volcanic soils. These and other influences mix and values fluctuate with discharge and even smaller spates. Standing waters are relatively stable, rivers show wide variations of mineral content. On the whole the conductivity of Nile waters varies between 150 and 300 μmho/cm; the low salinities of the Amazon are unknown in the main Nile system.

The reaction of Nile water is alkaline, with pH generally at 7–9, lower values exist. Oxygen conditions are generally favourable, deficiencies occur inside swamps and fleetingly in bottom layers of lake Victoria and in some dam-lakes. Supersaturations under the influence of algal blooms are frequent. Nutrients throughout the whole area are not abundant and some may limit plankton production at times, but on the whole Nile water is a favourable milieu for life.

Biology

Different habitats with different biological development follow each other along the Nile system. The two head lakes, Victoria and Tana, are self-sufficient systems with a circulation of nutrients and a full development of all interconnected lacustrine components. Such close connection between the components of aquatic life does not exist in the river as a whole; this is discussed in Part II. The biological exploration of the Nile system is uneven, some important habitats such as the waterfalls, the rocky stretches and the cateracts are little known. Not much known is the benthos, only some parts have been examined and some groups of animals. This incompleteness is due to the difficulties of work on such a long river and the remoteness of sites.

The best explored features are Lake Victoria and its general limnology and phytoplankton; further the Upper Nile swamps, their vegetation, plankton and fauna. The hydrology of the river system all along its course is well known, though changes have occurred due to the dams. Two biological components of life link the whole river together: fishes and the floating community. The fish fauna of the Nile is known and continuous except for the two head lakes. Lake Victoria is famous for its 'speciation' of some fish groups; this population is isolated from the river fauna by the Murchison Falls. In much lesser degree such separation has happened with Lake Tana. But the fishes of the Aswan High Dam basin are the same as in the rest of the river thousands of kilometers upstream and so is plankton. The most important binding factor is the water, which has come to Cairo from Uganda and Ethiopia.

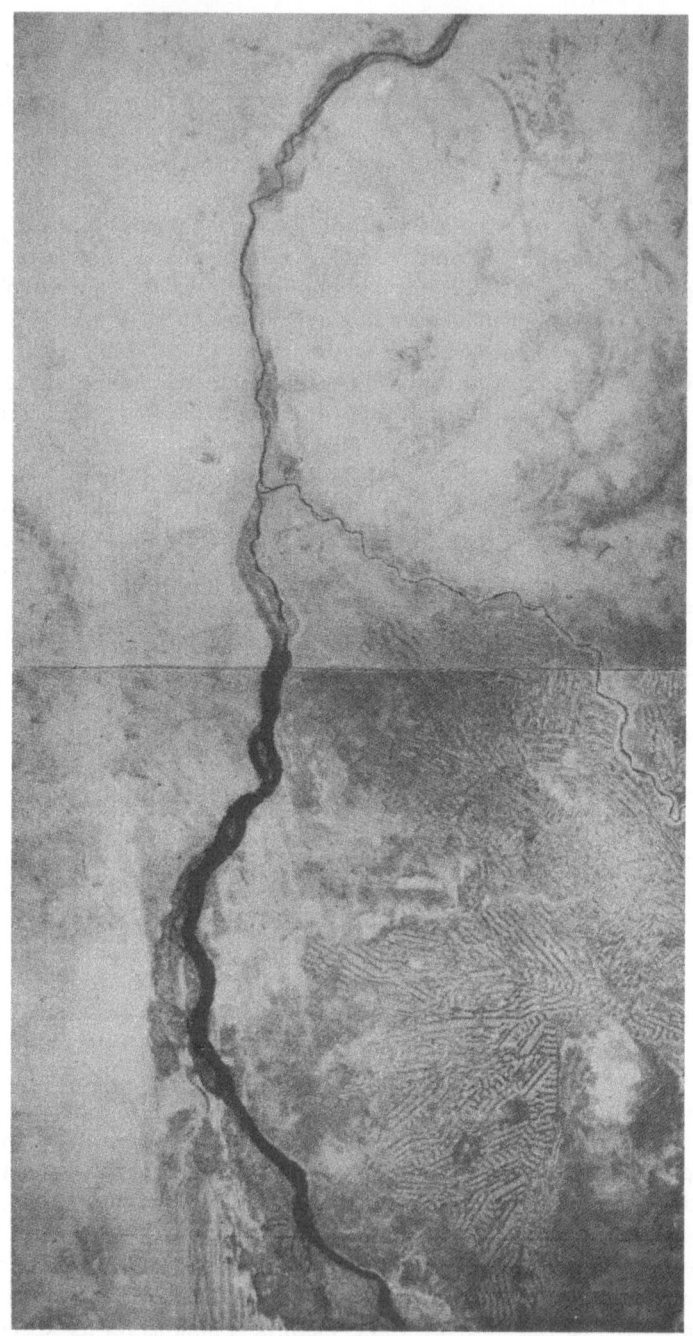

Fig. 20. The junction of the two Niles at Khartoum from space. Note the difference between the straight course of the White Nile, in its broad valley, and the winding course of the Blue Nile at low level. At the time of the photo the Gebel Aulia dam on the White Nile was full, creating a long river-lake. Between the two rivers the dark network of irrigation canals of the Gezira district can be seen. To the north the twist of the joint river represents the Sabaloka gorge, called the 6th cataract. The panorama embraces about 350 kilometers. ERTS.

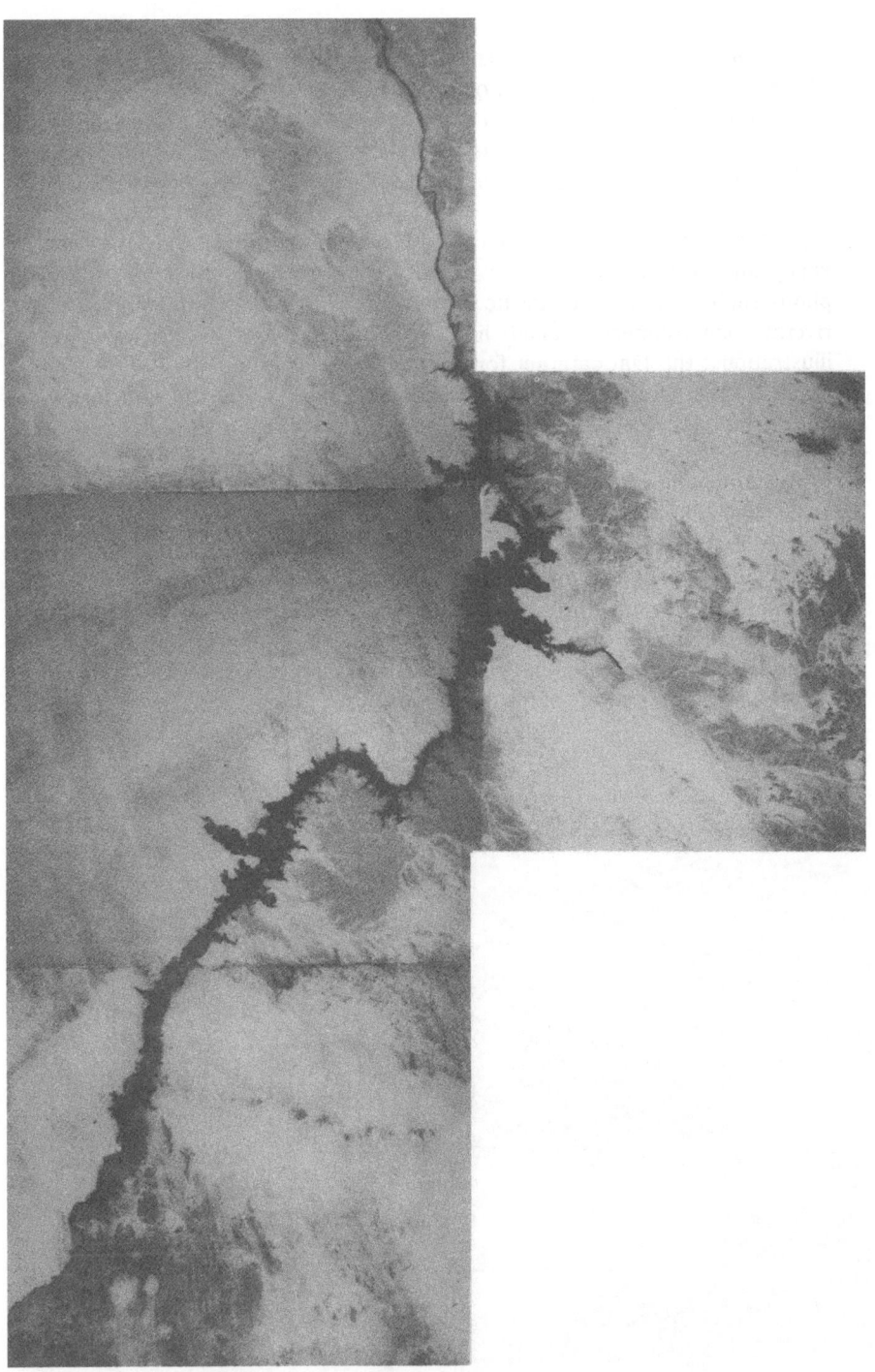

Fig. 21. The Aswan High Dam basin from space. Picture taken in June 1972, when the basin was only half filled. Note the dentritic expansions, denoting ancient drainage channels or rivers. The view contains the new lake from Aswan to the southern reaches, about 400 · kilometers. ERTS.

The 'life artery' remains, the landscapes change and their contents shrink. A combined space photograph embraces the junction of the two Niles in a semi-desert, the broad expanse of the Gebel Aulia reservoir on the straight White Nile contrasts with the meandering incised Blue Nile and its ancient creation, the fertile Gezira plain with its irrigation network. (Fig. 20.)

Further north the Aswan High Dam basin has drowned part of the river valley and created a totally new environment in sheer desert; the space photograph reveals the dendritic inlets, traces of extinct rivers. What this riverain environment looked like before the dam, is seen in two illustrations: the Dal cataract forms the end of lake Nubia-Nasser; enormous rock boulders, the narrow fringe of Acacia woodland and in the background the desert create a habitat not expected in a river of flat plains. (Figs. 21 & 22.)

The drowning of the river valley forced the exodus of the Nubian population fiercely attached to their ancient homeland. Their small patches of riverain crops have vanished, their date palms are gradually submerged by the advancing flood in 1964–5. The bare desert remains. (Fig. 24.)

The end result is seen in the almost lunar landscape of the new desert shores of the newly created lake. Desert rocks and dunes form the surroundings, the population has gone but new fishermen's camps have been set up, to reap the harvest of the lake. (Fig. 23.)

Fig. 22. The Dal cataract, 480 kilometers from Aswan to the south. This will be the end of lake Nubia, as the southern part of the Aswan basin is called. Note the boulders and the ripple marks left by the river. From colour photograph by B. Entz.

Fig. 23. The newly created shore of the Aswan High Dam basin consists of the desert dunes descending straight into the lake. The photograph illustrates the Arminna region about 250 kilometers south of the dam. From colour photo by B. Entz.

In Egypt landscapes are restricted to desert and alluvial river valley as seen in Fig. 10, broadening into the great and famous cultivated areas of the Fayium and the delta (Figs. 11 and 19). The four coastal lakes in the delta, Maryut, Edku, Borrullous and Menzalah are just descernible in Fig. 19; brackish and marine species of fishes and invertebrates mix there in a complex way. Otherwise the whole of Egypt is transformed by man in the last 6,000 years. A magnificent documentation of this process exists in the murals, papyri and even hieroglyphic signs of the ancient civilisation.

Fig. 24. The flooding of Nubia (1969) and exodus of the population before the advancing water of the Aswan High Dam basin. Note gradual drowning of the date palms and the desert on the opposite side. From colour photograph By B. Entz.

Human presence, exploration and pressure

Prehistoric evidence of man's presence in the Nile valley has been established at present for the last 50,000 years; archaeologists supply new discoveries continuously. No river in the world has attracted so much and almost passionate interest. Its geographical exploration, deep into the African continent on the other hand, is relatively new. The last links in geographic coherence of the river system were established only 100 years ago.

Human pressure on the lands of the Nile reach their maximum in the delta, where 600 people live and eke out their existence in every square kilometer. Further south extend the almost empty spaces of the desert; human densities increase in the middle and southern Sudan, Ethiopia and above all in Uganda. Around lake Victoria 50 people per square kilometer have been recorded, one of the highest densities in rural Africa.

Case 2. The Zaire River

Location and dimensions

The Zaire river, formerly called Congo, lies between lat. 13° S and 8° N in
the central part of Africa and its basin occupies about ¼ of this continent.
The basin covers 3,822 thousand sq. kilometers, the length of the main
river course is 4,650 kilometers; an enormous net of tributaries enters the
main river. The upper course of the Zaire is formed by the Lualaba river in
the east. A map shows some of the main affluents forming three groups
from the north, the east and the south (see Fig. 1).

Origin, geology, morphology

According to Robert (1942, Fig. 4 and Tables p. 29 and 31), a large
alluvial central basin is of Pliocene–Pleistocene age, surrounded almost
concentrically by nearer layers of Carboniferous, Permian and Triassic
formations; these in turn are encircled by crystalline Pre-Cambrian rocks.
Thus the longer tributaries pass through zones of different geological
character. They descend from elevations of up to 3,000 meters in the east,
falling to 1,500 meters and less in the north and south.

The slope of the main river, in its central part, is not great; over the last
1,600 kilometers it falls from 450 meters to 30 meters near the delta. But
this slope is not uniform; even stretches are interrupted by extensive
cascades and two waterfalls.

The central basin of the Zaire was filled in the Pliocene by a great lake,
which broke through a coastal rock barrier to the Atlantic ocean (Robert
1942). It seems certain that the present river system is composed of parts,
joined by 'capture' especially of the Lualaba and possibly other rivers.

Regime of Zaire river system

This is a river of water abundance, it discharges about 40,000 m³ sec. into
the sea. Rainfall of ca. 2,000 mm per year falls in two main periods but
tends to unimodality towards the north, like in the Nile basin. Floods are
of moderate 3–4 meters height, buffered by extensive swamp depressions,
riverain expansions, standing waters of various dimensions and inundated
forests. The Zaire river system is characterised by extensive rapids in three
main groups, each of some hundreds of kilometers long. Two of these are
on the Lualaba extension of the lower Zaire, one is called appropriately
'Portes d'Enfer' (gates of hell). The third is between the capital Kinshasa
on the Malebo (Stanley) Pool and the coastal town of Matadi, with a fall of
270 meters in 340 kilometers. Width and depth of the river varies greatly
from broad expanses over 12 kilometers wide to 800 meters, hemmed in by

rocky narrows; depth may attain 25 meters and dwindle down to shallows of few meters. Current velocities vary accordingly from below 0.1 m and lower in standing water stretches to 2.0 m per second. Such power of water movement causes the shifting of the bed load, islands and sandbanks are formed and eroded. Such conditions of quiet and turbulent stretches are repeated along the river system, defying 'normally' accepted zonation concepts.

Climate and vegetation

A tropical-humid climate prevails over most the Zaire basin. Air temperatures fluctuate little over the year between 25 and 30°C. Although soils in much of Africa are poor, large parts of the basin are covered by forest, now reduced by human pressure to 34% of the area. Extensive cultivations have replaced the original vegetation. In the river valley papyrus and other emergent plants fringe quiet parts of the rivers and interspersed or adjacent standing waters in the central basin of the Zaire and some stretches of the Lualaba. In pools and swamps a rich floating and submerged vegetation develops, but little is known about this. Notoriety has been accorded to the calamitous invasion of *Eichhornia crassipes* which has entered the Zaire some decades ago and has seriously affected navigation and the biology of the basin. It is almost certain that the invasion of the water hyacinth into the White Nile, around 1957 occurred across the Zaire–Nile watershed.

Limnology, water characteristics

The limnology of the Zaire system is not known adequately. The great efforts of the Belgian hydrobiologists were concentrated on the exploration of the lakes of the Western Rift valley and on fisheries and fish culture. Several institutes were active in these fields. The river system was only explored in fragments. These are the sources of the following general remarks.

With a great variety of soils and vegetation traversed by the network of rivers, water characteristics vary also. Temperatures are equably warm, except in tributaries at high altitudes where 13°C have been recorded; in the plains and in most of the central basin water temperatures are around 25°C all the year round. The colour of waters is dark in forest rivers, turbid in tributaries and the main river with strong and eroding currents and sediment transport. Sediments fall out with current loss and many quiet stretches and standing waters are clear. The pH shows a great variety. A. Berg (1959, 1961) has studied the impact of the reaction of many river waters and their mineral contents on the distribution of the water-hyacinth (*Eichhornia crassipes*). The invasion of this plant occurred about 1955, but probably earlier; it spread with grave consequences to navigation and other human activities. In our context Berg's survey of Zaire waters is of great value. He divides them into 3 major groups: 1. Humic acid waters, in

36

which acidity of between 3.5 and 5.2 is caused by the decomposition of vegetal matter in the swamp forests of the central basin of the Zaire; this is coupled with zero alkalinity and very low ionic content. 2. Humic waters partially neutralised, generally located in the central basin, but rising in the higher plateau regions; pH is from 5 to 7. 3. Humic waters completely neutralised; these are located in the plateau regions or at the beginning of their course through the central basin. To this category belong also the rivers of the eastern prolongation of the Zaire river, the Lualaba system. *Eichhornia* does not thrive in waters below pH of 4.2.

In addition to the work by Symoens (1968), Marlier (1973) and de Bont (1956) the studies by Berg allow us to have some general but not detailed impression of the milieu of Zaire water. Its mineral constituents are poor in the central part, richer in the eastern part of the system; it is astonishing that the water-hyacinth found all the opportunities to flourish. de Bont (1969) has reviewed 'The status of limnological knowledge of the Congo basin' with a considerable bibliography.

Biology of the Zaire system

The great variety of waters, their depth, width and currents, creates many different biotopes or places of distinct life-conditions. Fishes are the best known indicators of this. About 660 species have been recorded from the waters of the Zaire. Poll has supplied a series of papers on the central basin and the Lualaba (1959, 1963, 1973), and Poll & Gosse (1963) have given a systematic study of the ichtiofauna. All these results have been incorporated into the excellent summary on the ecology of the Zaire system by R. H. Lowe-McConnell (1975). She quotes Poll (1957), who has divided the river system into 5 regions, each with some characteristic species: 1. the lower Zaire with euryhaline and some freshwater fishes; 2. the region of 32 rapids extending 340 kilometers between the first region and the Malebo (Stanley) Pool; 3. that Pool itself with an area of 500 sq. kilometers; 4. the central basin with the greatest number of species; 5. the Upper Zaire or Lualaba system with nilotic and other affinities. This division reflects the general ecological differentiation of the whole system. R. H. Lowe-McConnell (1975) gives in Appendix 4 the number of species recorded in the various investigated parts; two examples of contrasts may be quoted. In the Malebo Pool live 122 species against 39 in the rapid sector downstream; 165 species have been found in the waters of the main river, but only 26 in the adjacent swamps. Adaptations to local conditions include fishes with adhesive suckers in the rapids, those with accessory breathing organs in the swamps, pelagic plankton feeders in open waters. The rapids, repeated in various parts of the whole river, act as barriers for some species, quite a different 'zonation' than that proposed for rivers of the temperate region. Finally it should be said that the Zaire fish fauna is 80% endemic and the richest in Africa.

The rest of the fauna is poorly known. Malier (1951, 1954) has investigated some of the high altitude courses of tributaries of the eastern region. These are typical 'rithron' streams in the definition of Illies. Conductivities are low, temperatures may be as low as 13°C, the usual assembly of Perlidae, Trichoptera, Ephemerida etc. live there and fishes and frogs with suckers. Sessile algae provide food for herbivores, net-building or sediment-catching insect larvae abound.

Parts of the Zaire with higher calcium contents have mollusk faunas, rapids abound in *Simulium*, both of great medical importance. Bilharzia and onchocerciasis are rife in addition to other tropical diseases.

It seems that plankton studies have been rare and confined to some standing waters and the plant component. An early study by van Oye (1926) was on a short tributary in western Zaire, the Ruki river. It runs through forest, has humic acids and carries an association of desmids, diatoms, chlorophyceae and rotifers. This is an incipient plankton which in all short running waters cannot develop to full capacity.

Human discovery, presence and pressure

Whatever happened in this part of the 'dark continent' is for students of African history to relate. To the outside world the first news about the mighty river was given by Portuguese discoverers in the late 15th century. Diego Cam embarked in 1482, saw the delta, ascended as far as the rapids at present Matadi and inscribed the Portuguese emblem on a rock. Other explorers followed, including the great Vasco de Gama. A series of penetrations followed, political and geographical, intensified in the 19th century, with H. M. Stanley in 1877 and P. S. de Brazza in 1880. This is the period of commercial and political activity of European nations leading to the carving up of Africa by spheres of interest and final possession. By the end of the 19th century the Belgians took over the 'Congo' and, with the awakening of African nationalisms, left in the 1950's. They left a legacy of considerable geological, mineralogical and hydrological exploration; the hydrobiological work on the river system has been summarised above. Their great work on the rift valley lakes is outside our scope.

Dams have been erected in the industrialised Katanga province for hydro-electric energy; Marlier (1964) has described some limnological effects, including the usual strong development of phytoplankton. Population pressure has made great inroads into the forest cover, large clearings for cultivation have been made; the average density of population is 4/sq. kilometer, concentrated in some parts.

Case 3. The Amazon System

Location and dimensions

The hydrographic basin of this greatest of all rivers extends to 6.5 million sq. kilometers and occupies more than a third of the continent of South America. The tributary network stretches from 18°S to 3°N latitudes. The length of the main river course is 6,770 kilometers, some 20 affluents are over 1,000 kilometers long. (see Fig. 2.)

Origin, geology, morphology

A geological map and two transects (in Sioli 1956) illustrate the main features. A central alluvial basin of Tertiary and Quaternary origin is bordered in the north and south by Palaeozoic and Archaic formations with smaller enclaves of Triassic and Cretaceous. In the Miocene the present basin was occupied by a vast lake or lakes and rivers, which drained into the Pacific ocean. With the rise of the Andes, this outlet was blocked and the new slopes forced the waters to its present flow into the Atlantic ocean. Although most of the rivers arise on higher grounds, yet only one/seventh of the whole river system is more than 200 meters above sea level. Gradients of river slopes flatten out to 1.5 and less centimeters per kilometer. Yet some of the tributaries arising in the Andes or their foothills carry large quantities of sediments arising from erosion; these are transported into the main courses and when deposited cause a continuous sequence of changes in the river bed and its shores. Lowering of the sea level during the ice age resulted in increased erosion and deepening of lower river courses up to 100 meters deep. Fittkau (1974) has given an excellent account of the geological history of Amazonia.

Water regime

The Amazon carries a water mass 4 or 5 times that of the Zaire and, incidentally 8–12 times that of the Mississippi. It discharges 200,000 m^3/sec into the sea. A rainfall of about 3,000 mm per year causes floods culminating in May and river levels may rise up to 20 meters. Large areas of the river valley are under water each year; this floodplain is called 'varzea' and is a characteristic feature of the Amazon. The very gentle slopes cause the rivers of the system to meander, braid, anastomose; they create islands, sandbanks, oxbows, mouthbays and internal deltas of bewildering complexity as seen in Fig. 26. No river in the world exhibits such intimate relationship between the land and running water and this finds its fitting recognition by the name 'Amazonia' given to the enormous area. There are no true lakes in Amazonia, all standing waters are created

39

by the river system.

The velocity of currents is dictated by slope, obstacle to flow and discharge. Obstacles of continuously created sediment banks reduce currents to imperceptible values in lagoons; on the other hand the watermass of the main collecting channel, the Amazon and its upper stretch, the Solimoes, forces currents up to 2 meters/sec. and to 3–4 meters/sec. in the narrow stretch at Obidos.

Climate, soils, plant cover, impact on rivers

Temperatures over Amazonia are intertropical – warm between 25°–30°C with a minimal amplitude of 2.8 degrees all the year round. The ample rainfall for many thousands of years, created the tropical rainforest which covers 84% or 4 million sq. kilometers, of the river basin. This is a climax forest on very poor soils and agriculture in clearances has not been successful except in parts of the 'varzea' floodplain, where sediments are brought in by the annual floods. This floodplain of 50,000 sq. kilometers is from 20 to 100 kilometers wide, flanked by firm land with the great forest. A transverse section of the floodplain reveals ridges of land dry during low water with gallery forest, standing waters, grassy depressions, river side arms; at high level of the river all these are under water for some months. (Sioli in Whitton et al. 1975, p. 472.) A map in the same contribution by Sioli (p. 476) gives the overall and longitudinal ecological division of the Amazon basin; a. the forest area enclosing the alluvial plain north and south, with poor soils and restricted fauna and flora; b. the Andean foreland encircling many tributaries in the west and their alluvial valleys to the main river; these form the soils of the 'varzea', rich in nutrients with optimal development of flora and fauna; c. the outer borders of the basin of mainly archaic, crystalline formations, soils poorer, flora moderately developed; d. strips of palaeozoic and cretaceous formations along the middle and lower alluvial basin with rich biological development. (p. 463, Sioli, in Whitton 1975.)

Limnology

Under the impact of soils, forest and geological regions 3 main types of rivers exist: a. 'White water' rivers, coloured by loamy sediments, turbid, mainly alkaline or neutral in pH; b. 'Black water' affluents, coloured brown by humic acids from forest soils, pH acid; c. 'Clear water' rivers from the south-eastern part of the lower basin, mainly neutral in reaction. All these affluents mix in the water of the main river, the Solimoes and its prolongation the Amazon, but these are predominantly of the white water type. The striking contrast of the black water rivers and the main river itself is illustrated by a magnificent air photograph of the rio Negro entering the Amazon. (Fig. 25.) A further distinctive feature of the Amazon are the 'mouthbays' of some of the lower tributaries as indicated in a map by Sioli (p. 468 in Whitton et al. 1975) and exemplified in Fig. 26 in the Tapajós

40

Fig. 25. Confluence of the Rio Negro and the Amazon at Santarem. Note the striking contrast of the dark, though sediment-poor, waters and the 'white' silt coloured waters of the two rivers. Note also the deposition of alluvial islands and banks by the action of the river. Location 1 on the map Fig. 2. From H. O'Reilly Sternberg, the *Amazon River of Brazil*.

river. These are interpreted as drowned river valleys due to the lowering of the Atlantic Ocean during the Pleistocene ice age.

A scanning-radar air photograph of the confluence of the Tapajos river with the Amazon gives an insight into the variety and complexity of the various river formations. (Fig. 26.)

Physical and chemical features

Water temperatures are high, up to 30°C and some times more, with almost no seasonal variation; transparencies by Secchi, vary between a minimum of 0.1 meter in some white water rivers to almost 5 meters in clear rivers; the black rivers vary between 1.3 to 2.3 meters. In the numerous permanent or temporary standing waters and mouthbays transparencies rise with the deposition of any sediments. The colour of the three types of rivers originate from the soils and rocky substrates of the regions traversed. White waters come from the erosion of mountain slopes, clear waters from the old massives of central Brazil and Guyana, black waters from regions of podzol soils with dissolved or colloidal humic substances. The electrical conductivity is almost everywhere below 100 μmho/cm, a sign of the low mineral content.

Investigation of the chemistry of Amazon waters started by the turn of the century and is continued up to the present. A table compiled by Sioli from many analyses (p. 474 in Whitton ed. 1975) gives a survey of the chemical composition of the river types and that of the main types of standing waters. The overall character is one of 'extreme purity' to use the expression of one of the first analysts. A detailed discussion is not necessary for the purpose of this essay. But the changes of water features which occur in rivers at one point are of great interest. Marlier (1973) quotes a paper by Oltman of the U.S. Geological Survey in 1966, who examined the water of the Amazon at Obidos on three days, about 400 kilometers from the delta. Unfortunately there is no indication of dates in Marlier's account. But even with these restrictions the data show that water of different composition has passed the observers point on the 3 occasions, an important point for river flow.

Table 2.
Table of water composition at Obidos, on three different days (Marlier 1973)

Elements	Values in parts/million		
Ca	4.3	10.0	3.0
Mg	1.1	0.4	0.6
Na	1.8	4.2	1.8
K	0.4	0.6	—
HCO_3	19.0	33.0	16.0
SO_4	3.0	6.4	1.0
Cl	1.9	4.5	1.6
SiO_2	7.0	9.0	7.0
Conductivity at 25°C	40.0	84.0	34.0
pH	6.5	7.1	6.5

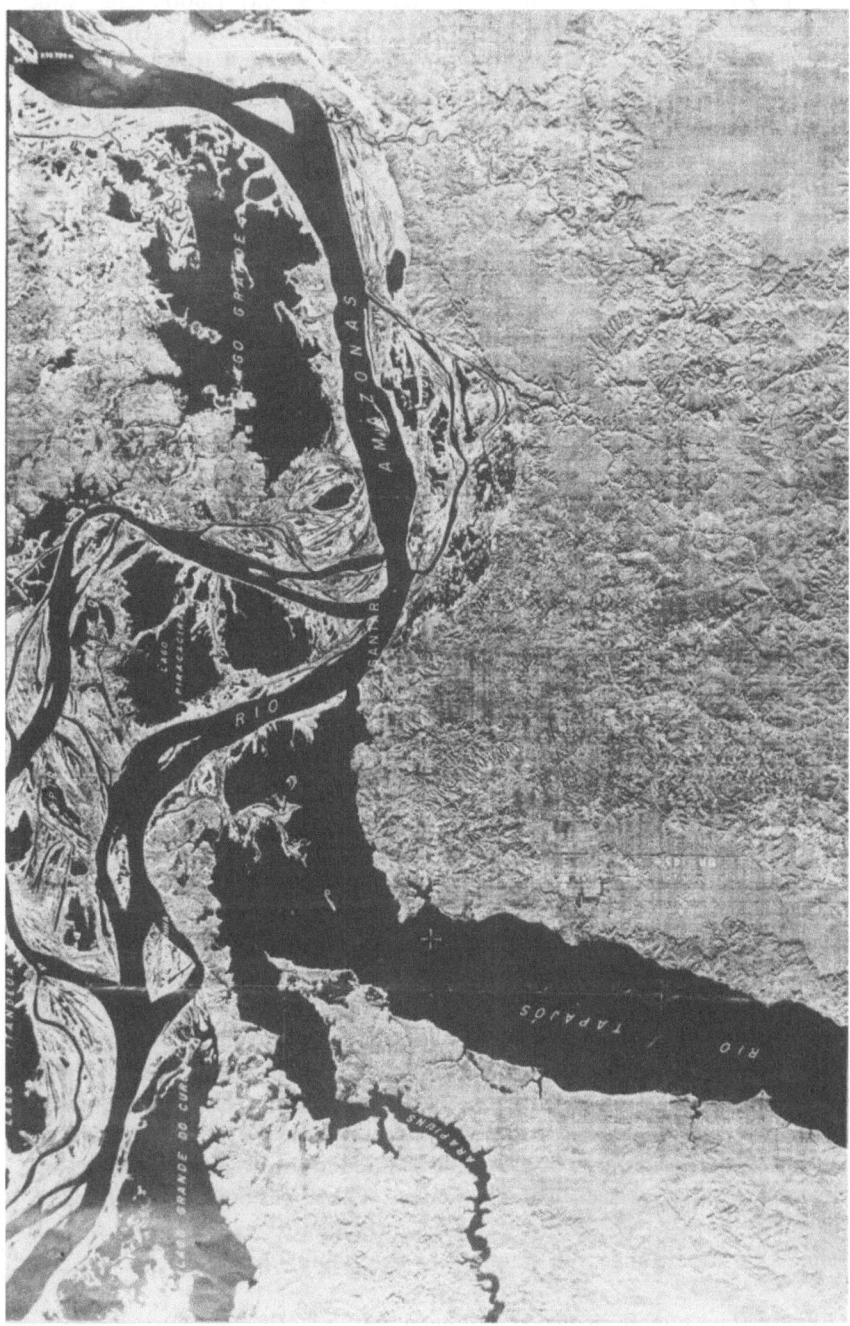

Fig. 26. Scanning radar photograph of the confluence of the clear-water Rio Tapajós and the Amazon. This picture gives a variety of Amazonian features: mouth bays, the braiding of river arms, standing waters built up by sediment sandbanks, the visible stowing effect of the main river. Photo by 'Projeto Radam, Rio de Janeiro'. Courtesy of H. Sioli.

The Amazon at Obidos flows through a gorge only 1,800 meters wide with near 100 m maximum depth and the current may reach up to 4 meters/sec. At this speed water sampled on the first day would have travelled in 24 hours 340 kilometers downstream and each subsequent sampling would be of different water, which is exactly what is shown in the analyses. This point is discussed in the second part of this essay.

Oxygen in Amazonian waters varies greatly; it may reach supersaturation in standing waters with algal blooms, but generally is present around 70% saturation.

Biology

Water vegetation reaches its peak in the 'floating meadows' developed in parts of the river system. Two main types exist, one composed of floating plants like *Pistia, Salvinia* and *Eichhornia* – which does not explode here to the alarming proportions as in the Zaire and upper Nile. The other type is composed of semi-terrestrial grasses *Paspalum, Oriza* and others, which together with *Carex* and *Scirpus* invade the surfaces of free water. Independent of water transparency these meadows occur in turbid waters; submerged vegetation can only develop in clear standing waters. Phytoplankton occurs wherever conditions of transparency and waterflow allow. In mouthbays of clear water rivers like the Tapajós a rich assembly of Cyanophytes, Chlorophytes, Desmids and Diatoms is present in depths down to 27 meters, but densest down to 10 meters; in the Amazon nearby this dwindles to traces. (Table p. 481, Sioli in Whitton et al. 1975.)

Animal life, where examined, varies greatly; the richest biotope is in floating meadows, where densities up to 700,000 invertebrates per sq. meter have been recorded. Crustacean plankton in clear waters may reach 140 individuals per litre. The bottom fauna is rich in the standing waters of the 'varzea'; so far 1,000 species of chironomids have been found. In these 'lakes', of up to 100 kilometers long and 40 wide, fishes thrive and bird life is abundant. Molluscs live in non-acid waters.

As usually fishes are the best explored group and, by their endemicity, are a testimony to a long development and a great variety of living conditions. About 2,000 species have been found in the whole of Amazonia, including adjacent Guyana, Colombia, Peru, Bolivia. This is the greatest and richest province of fish distribution in the world. The 1,300 species of the Amazon proper are not evenly distributed in their density; fisheries are poor in dark rivers, richer in white rivers. The upper links of the foodweb seem to be the 4 species of caymans, now almost exterminated with ill effects on the balance and standard of fish populations.

Exploration, human presence and pressure

Above I tried to convey an impression of this gigantic river system gained from some existing sources. Only fragments could be pieced together selected for the purpose of this essay. The exploration of Amazonia is continued vigorously by the cooperative efforts of the Instituto Nacional de Pesquisas da Amazonia and the Max-Planck Institute for Tropical Ecology at Plön in Germany; the latter's director, H. Sioli, has provided a series of informative general surveys of the Amazon system. The drive and scope of research is best illustrated by a list of publications issued from Plön in 1976. Of about 300 entries, 60 are devoted to soils and forest, 170 to various problems of aquatic biology, the rest to human impact, conservation and books. Two volumes have appeared of the 'Biogeography and Ecology of South America' in 1968, 1969; a periodical 'Amazoniana' is in its sixth year of publication founded by Brazilian and German institutions and published in Germany. It contains a stream of papers on the Amazon.

Man in the Americas stems, according to archaeologists, from Asian immigration via the Bering Strait during early intervals in the ice age. It is suggested that man arrived at the southern reaches of South America around 10,000 B.C. Be this as it may, archaeological evidence is gradually forthcoming. Some radio-carbon dates on pottery from a river site on the Careiro island gave dates of 2050 B.P. but may be older. (H. O'Reily-Sternberg 1960.) Spanish and Portuguese explorers penetrated Amazonia in the second half of the 16th century. Journals with descriptions of the country and its Indian inhabitants were kept; the first map was drawn in 1637 (Sioli 1967). Alexander von Humboldt in his 'Voyage aux régions du nouveau continent' (German edition 1859) wrote enthusiastically and with understanding on many features of the Amazon. Maps improved greatly in the last decades through photogrammetric and radar-space photographic methods.

With increasing penetration by Europeans, the indigenous population dwindled. Agricultural schemes started and through misunderstanding of the poverty of cleared forest soils many failed. Varzea soils gave better results. Sioli (in Meggers et al. 1973) has mentioned the population densities in the state of Amazonas at 0.3/sq. kilometer and 0.9 in the state of Pará. The population is thin outside the main towns of Belem and Manaus, rivers are still the main communication lines; 40,000 kilometers are navigable, ocean going vessels can penetrate as deep as Manaus. There is no pollution, no dams exist. The construction of the new highway is destroying forests at a large scale. Tropical diseases are widespread.

Nile, Zaire and Amazon, comparison and conclusions

I have tried to summarise the main features of three of the greatest rivers in the world, in order to find the basic nature of long, running waters. The treatment of the three rivers has been uneven because of personal involvement in the Nile and its great human implication and documentation.

The three examples of long rivers *differ* in many respects. The Nile has undergone drastic climatic changes over much of its basin, it has become a desert river in the north; the Zaire and Amazon lie entirely in the warm-humid equatorial zone. This has profoundly affected their water supplies and the vegetation cover over their environments. The maps of their systems (Figs. 1 and 2) show the differences in the tributary systems, the Nile is poor when compared with the abundance of Zaire and Amazon. Although all three rivers have enormous drainage areas, this is theoretical in the case of the Nile, where 44% contributes nothing. Unlike the two others, the Nile is a managed river now, its use dates back 5,000 years. No words or statistics illustrate the contrast between the untamed Amazon and the Nile better than Fig. 26 and the colour plate (Fig. 27), where water scarcity, main supply and its use is shown.

But their *common* features are of interest in this essay. All are long rivers, from 4,650 kilometers of the Zaire, to 5,600 for the main axis of the Nile and 6,700 kilometers for the main course of the Amazon. They dominate the plains through which they flow by transporting water and sediments, they build and change their valleys by the force of their motion. They change their courses thus creating a variety of standing and running waters with different conditions for life.

These habitats may be arranged mainly along the river, or to use Gessner's (1955) excellent expression, like beads on a string; or are arranged also horizontally across the river valley or flood plain. The full potentials of aquatic life, that is water vegetation, plankton, shore- and bottom fauna, are present in all three rivers. They will appear where conditions allow, though sometimes not simultaneously. Above all the water medium in rivers moves continuously in space and most habitats cannot share their metabolic processes as in lakes.

Human presence, pressure and scientific exploration

Man's influence on the three rivers is so uneven, that we must refer to the appropriate chapters of each river; so has exploration.

Scientific investigations on the three rivers are also unevenly advanced. The Zaire is not well explored, its limnology is known only in fragments, from which I tried to piece together some general story. The Nile has a bibliography, which could fill a library but its limnological or better biological investigation has still many gaps, in spite of the recent monograph (1976).

On the Amazon work is continuous. Two research institutes, one Brasilian and one German, are in close and successful cooperation and a special periodical is devoted to the long-term work on this greatest of all rivers.

Fig. 27. Three colour photographs depicting the essence of the Nile: The scarcity of water in the desert north of Khartoum; after an occasional rain, some small drainage run-offs may contain some water under the surface. The junction of the two Niles from the air. The contrast of the grey water of the White Nile and the dark green water of the Blue Nile is evident. A low level and the blocking effect of the White Nile cause a drastic fall in current and an algal bloom in the other river; a lesson in limnology! Both pictures by Kuhn. Airview of the fringe of cultivation in the Egyptian delta: irrigation versus desert. Note the sand dunes; the irrigation canals are fed with water from 4,000 kilometers to the south; a lesson in river transport! Photo by J. Talling.

46

Part II.

General and Polemical reflections on the nature of rivers

'What is a river'?

Under this title a number of papers was delivered at an American symposium a few years ago (Oglesby et al. 1973). Subjects varied between hydraulics and distribution of plants in rivers and occupy 74 pages of print. The chairman of the session tried to summarise the rather disjointed opinions expressed by saying that 'aquatic life in rivers forms a loosely interacting community because of the downstream drift and the migration upstream of fish and invertebrates ...'. In the same session an expert on hydraulics gave his definition of rivers as 'concepts resting on geomorphological theories of minimum variance and minimisation of work, simplified for biologists by the statement, that one cannot both maximise both efficiency and stability in biological systems'. This was declared in the final session of the symposium to be a 'definition of river systems in their most fundamental way'.

This is probably profound and presents maybe one of the new formulations of theoretical ecology. For me it is too sophisticated and such formulation shows also the lack of mental 'bridges' between biology and the exact science of water motion. The obvious solution is that biologists (limnologists) should base their work on an inter-disciplinary approach; on the other hand specialists studying hydraulics could learn from the reactions of organisms.

From complex terms and mathematical concepts let us turn to the common sense of ordinary language. After all, language represents the experience of man accumulated since the beginnings of ideas. In every language words have been in use for a variety of running waters; in English there are 'brook, beck, rivulet, stream and river'. The French have 'ruisselet, ruisseau, rivière and fleuve'; the Germans have separate words for 'Rinnsal, Bach, Fluss with diminutives, and Strom', which incidentally denotes a large river. If you turn these expressions on your mental palate, you will probably discern a progression in size of running waters. In later pages I shall use the term 'ontogeny'.

Tacit acknowledgement to this succession of sizes (and life conditions) has been made by scientists in attempts at zonation in running waters. These were made mainly in temperate regions and based on the observed distribution of invertebrates, vegetation types and fishes. Classification attempts of running waters included also temperatures, chemical composition of the water, pH and 'trophic' standards. One eminent limnologist used 13 variables to embrace any possible diversity, which would create an enormous scheme. Illies and Botosaneanu (1963) have made a survey of the problems and methods of classifying running waters on the basis of their biology. They assembled a large volume of references up to 1961, confined mostly to the then fashionable upper reaches of running waters. The authors restrict themselves to the steep and turbulent stretches, the 'rhitron' zone, in contrast to the quieter, downstream zone of the 'potamon'. The authors of this scheme claim its validity to all regions of the world and indeed such succession exists in the fauna of many parts.

Illies & Botosaneanu do not deal with the 'potamon', I am not concerned with the 'rhitron', nor with the ecology of 'streams' but with the characteristics of fully developed rivers. These have been neglected in hydrobiology; it is striking that Hutchinson's great work contains only few remarks on rivers and during the last congress of the S.I.L. (International Association for Limnology) only 10% of the communications were on running waters and most of these again on short streams and the rest on pollution problems. Admittedly practical issues of river-management are pressing in industrialised societies and few rivers are not affected by human impact.

Books on 'river ecology' recently published reflect this restriction and produce therefore a narrow and sometimes erroneous survey of the nature of rivers. In earlier work a curious discrepancy appears; some investigations gave excellent results, others were apparently not aware of them. This is illustrated by the case of river plankton, an important indicator of water flow. In Britain, Carpenter (1928) denied the existence of 'potamoplankton'; Butcher (1940) declared it to be 'but a pale image of the benthos'. Both seem to have based their opinions on British 'rivers', which I regard as 'juvenile' stages of running waters, because they are short. Yet at the time of their writing phytoplankton was found in the Thames by Fritch, and in 1928 the Volga monograph had appeared.

More serious is the curious opinion, voiced quite recently (in Oglesby et al. 1973). Discussing the biological components of 'What is a river?' one contributor said: . . . 'Although large rivers, particularly because of the extensive interference with unrestricted flow, contain a biota that may be described as planktonic, these organisms are simply in transport and at least the animal portion can best be described as drift.' This, I am affraid to say, shows a lack of knowledge of the potential biological creativity of water. But even in usually well informed recent books on river ecology the appearance of plankton is not represented adequately. Hynes (1970) knows Russian sources on zooplankton, yet he regards . . . 'crustaceans as rarely numerous in the open water of rivers'. Those found besides *Cyclops* and *Bosmina. Alona, Chydorus* and *Diaptomus*, other genera of Cladocera, are unusual and are 'strays from other habitats'. More justice is done to the pelagic component of river water in the book, edited by Whitton (1975); plankton is recognised as depending on the 'age' of water and as integral part of the lotic community. The gradual appearance of members of this community is mentioned, first diatoms and rotifers. This sequence will be commented upon again later on, but it seems to indicate that the author of this particular chapter has had limited experience with long rivers.

Review of opinions and the neglect of older results.

It seems that scientific literature written previous to the modern escalation

52

of hydrobiology is sadly neglected. At the turn of this century, when inland waters became a worthy object of study, the question was raised whether rivers had a 'potamoplankton'. The search was widespread over Europe, Russia and America.

First must be mentioned the pioneer work in the United States, because of the freshness of approach and its perseverance. C. A. Kofoid carried out his 'Plankton Studies' on the Illinois river from 1897 to 1910. This work was continued by Forbes & Richardson up to 1919. The existence of river plankton was established, the important factors governing its appearance, the 'age' of the water flowing, were recognised as a fundamental; incidentally also by German hydrobiologists. Forbes & Richardson (1919, p. 147) arrived at the remarkable formulation: 'A river and its plankton are a flowing soil and its crop, both slipping away continuously, but both renewed constantly from an exhaustless source of supply. . .'. On p. 148 we find: 'The river being continuously fertilized . . . if it has a slow current, as in low water and a long course, . . . Its plankton may multiply . . .'. In 1928 Forbes surveyed the many years of work on the Illinois river with 20 bulletins and 2,000 pages.

Others followed; Galtsoff (1924) examined 405 miles of the Mississippi with reservoirs interspersed. He observed the fall-out of sediments in his 'lakes', the clearance of the water, the increase and later decrease of transparency due to a vigorous development of plankton. On p. 398 he says: 'Velocity of current is the principal factor that affects life in the river'. Rapids destroy most of the 'microscopic life' but upstream there is an increase. He notes the inconstancy of river plankton in densities and composition, even over short distances. Coker observed that the Keokuk dam on the Mississippi built in 1913, prolonged the 'age' of water and increased the biological production including fishes 30 to 50 fold. Eddy (1931) noted the gradual appearance of plankton, first diatoms and rotifers; he was aware of German work and Behning's studies on the Volga.

At the same time as this early work in America was under way, European biologists were busy in several countries. In Germany the dispute on life in running water was prominent and a number of rivers were studied. Russia had, of course, a much wider span of rivers, as their length grows with the area of the land. The greatest river of Europe, the Volga with 3,600 kilometers in length, attracted much attention for many years. This reached its culminating stage in Behning's monograph 'Das Leben der Volga' published in 1928 and in a summarising paper a year later (1929). Behning recognised the length as important for a full development of life. He says that not all what floats in a running water is plankton; a true 'potamoplankton' can only develop where a long course or a low velocity of current allows for it. The zooplankton of the Volga comprised 2 cyclopids, 3 calanoids, 15 cladocerans. Other rivers of European and Asian Russia were also investigated. A rich literature both 'old' and recent exists on rivers of the whole Soviet Union. Many rivers have been dammed and a number of investigations on the first years of biological events in these dam-lakes made. I have tried to summarise some of the publications in a chapter of the book on 'Man-Made Lakes' edited by R. H. McConnell (Academic Press 1966). But a survey of the many results on rivers in Soviet Asia should be made available to the scientific world at large. A new monograph of the Volga, now completely managed by several dams would also be desirable. I owe Dr. N. N. Smirnov gratitude for supplying me with some new references.

Of the numerous publications of that period, the general review of studies of large rivers by J. des Cilleuls (1928) must be noted. He collected references then available from Europe, Asia, Africa and America and noted specifically the sequence of the appearance of plankton components, diatoms and rotifers.

Monographic studies of rivers are few; the Illinois work for many years is one. R. Lauterborn (1916–1918) gave us an outstanding account of the geographic and biological sequence ('Gliederung') of the Rhine. For its time this was an achievement, describing the succession of landscapes, the reduction of currents, the creation of adjacent bays and the development of life communities including plankton. Then came the monograph on the Volga. After a long interval Liepolt's limnology of the Danube (1967) appeared, and in 1976 I tried to write, collate and edit the monograph on the Nile, a multidisciplinary treatment of a river and valley, its past and present state.

All these rivers have now been changed by human action. But except the last two publications, all above contributions to the ecology of rivers have appeared before the thirties of this century, in the first phase of hydrobiological exploration of inland waters. They have been forgotten and largely ignored in the present treatment of rivers. Even the excellent remarks by Gessner (1955) seem to have had no impact, yet his chapter on 'Die ökologischen Wirkungen des strömenden Wassers' should be read by everybody dealing with running water. It is an astonishing proof of the deficiency of some modern scientists; the very basis of science, of careful scrutiny of achieved results, has been neglected and the lack of knowledge of languages has produced deplorable effects.

Factors governing events in rivers, currents, water transport

The factors governing events in rivers were already broadly recognised by early observers: water, in unidirectional motion, its speed, its action of transporting its suspended matter. Time of flow has been named as the 'age' of water and this depends on the length of a river; it varies with obstacles to flow which can be natural or man made. Within the constraints of these governing factors biological events take place.

All this has already been said and has apparently been forgotten; I claim the merit of bringing these early achievements to the surface of the present. But a more precise documentation is necessary to two fundamental features of rivers, the motion of water and its transport.

Current speed varies horizontally across a river, vertically in the moving water mass and longitudinally with discharge, bed morphology and

54

landscape. For horizontal differentiation an example is presented from the swamp region of the upper Nile.

Table 3.

Current velocities across a vegetation fringed river; in meters/second

Depth in meters	Inside the vegetation fringe	1 meter outside	midstream
0.2	0–0.2	0.32	0.54
1.0	0.1	0.46	0.72
2.5	—	0.39	0.80
3.0	—	—	0.70
3.5	—	0.35	0.59
5.0	—	0.29	0.29

Similar results have been obtained by Migahid in 1948 who studied the influence of water flow on the distribution of the swamp vegetation. We see that the vegetation fringe provides shelter for the rich assembly of organisms living there. The effects of obstacles to flow have been studied in the basin of the Gebel Aulia dam in 1953.

Table 4.

Slow down of currents upstream of a dam; in meters/second

Distance from dam in kilometers	Mean velocity	Vertical amplitude of current
320	0.35	0.25–0.42
224	0.18	0.14–0.19
206	0.15	0.11–0.17
94	0.08	0.08–0.12
2 from dam	0.07	0–0.12

The effects of the loss of currents are sediment shedding upstream, clearance of water and rich plankton production.

Finally the transport of water passing an observer's point is demonstrated by calculations at different speeds.

Table 5.

Travel of water past an observer's point at different speeds in meters/sec, one and 24 hours

Current velocity	Distance travelled in one hour	Distance travelled in 24 hrs.
m/sec	meters	meters
0.1	360	8,640
0.3	1,080	25,920
0.6	2,160	51,240
1.0	3,640	86,400
2.0	7,280	172,800

Expressed in words the water mass sampled for some reason at one point has moved away and a new water mass faces the observer. This is true, but we have already mentioned that river flow is very differentiated both horizontally and vertically. Besides, morphological features influence the flow, such as meanders (leading to oxbows), braiding, narrowing of river channel, rocky sills causing cataracts or cascades, and the hold-up of flow of one river on a tributary. Abundant examples exist especially in the Amazon, but also in the Zaire and Nile. Figures no. 13, 14, 16, 22, 25, 26 and the middle of the colour plate Fig. 27, offer examples.

The best method to measure the travel of a water mass is by release of colour indicators. This is repeated in nature by the brown flood water of the Blue Nile. It arrived at Khartoum in 1967 on the 15 of June and advanced to the Wadi Halfa end of the rising Aswan High Dam basin, 1,500 kilometers away, on the 27 June, in 12 days. The basin was then only partly filled, now the travel of the floodwaters advances much slower. In the hydrological documentation a number of observations exist on the travel of 'hydrological fluctuations' or disturbances. These observations were of importance before the dams were built, now all this has changed. Entz (in Rzóska 1976) has used conductivity measurements of the laminar water masses spreading over the Aswan High Dam basin; he concluded 'that during the period in question (1970, 1973–1974) it took about six months for the front of the flood to pass through Lake Nasser ... to the High Dam.'

The second important function of a river is the *transport of water*[1] and its contents. The colour space photograph, which shows the Nile in the desert is the most convincing documentation. The sediments carried during the flood of the Nile, have created the alluvial soils of agricultural importance both in the Sudan but above all in Egypt (see Figures 10, 11, 19, 20).

Einsele (1957) has studied the carrying capacity of currents in some Austrian rivers. At a speed of 1.70 m/sec and above scouring of the river bed occurs, even bedload and gravel are moved. A gradual decrease of the hydraulic force takes place, at a velocity of 0.5 m/sec sands fall out, at 0.2 fine particles begin to settle out. Einsele comments on the effects on fisheries. There is ample evidence of this fundamental effect of water motion in the Nile basin; the latest example is the deposition of sediments at the southern end of the Aswan High Dam basin (lake Nubia). A process lasting for many thousands of years, depositing the Nile sediments over Egypt and part of the sea, has been totally changed by man. Not so in the Amazon. There the force of the river system is untamed and continuous shifts of sediments modify the river valley profoundly as seen in Figs. 25 and 26.

Fragmentation and Potential of life in rivers

Water in flow has profound biological effects. Ambühl (1959) studied

1. An American specialist on hydraulics asserted to ecologists that 'the main task of a river is to conduct water ... not Sunday boating, fishing, disposal of old tyres ... nor ... to support life and its ecology. Every student of ecology should familiarise himself with the rules and laws of hydraulics, of which the motion of sediment is an important factor'. (Oglesby et al. ed. 1973, p. 318.)

these effects in Swiss streams. Rocks of various sizes cause hydraulic roughnesses, which result in turbulent flow with vortices, eddies and 'dead' areas. A highly adapted fauna of insect larvae (and others) live, feed and find shelter in the various areas.

Gessner (1955) has contributed greatly to an understanding of conditions in running waters. He says that the only unity of a river lies in its name; landscapes, biotic and abiotic characters change along the river course, biotopes follow each other like 'pearls on a string' (pp. 285–6). The water medium is constantly replaced, which affects the floating association, less the organisms of the substrate. But even vegetation shows a succession both longitudinally and seasonally (p. 287). Current velocities, changing in time and space create a mosaic of conditions, to which plants react both morphologically and in their physiology (pp. 293, 299). Gessner deals with the effect of river flow on the Amazonian floodplain (pp. 317–320). A special chapter (pp. 461–479) deals with the plankton of rivers; he notes the early discussions on 'potamoplankton', its continuous supply and removal; the faster the current the less plankton can exist. The seasonal distribution of plankton in temperate rivers is replaced by conditions of seasonal turbidity, which affects the photosynthesis of phytoplankton.

Gessner quotes A. von Humboldt that unity in multiplicity is nowhere as clear as in the 'tropical river, where flowing and standing waters, climate and landscape, the present and the past and the geological history form together with the organic world the great unity of nature'.

In the upper White Nile a swamp region extends for 600 kilometers after the descent from the Ugandan escarpment. A wall of *Papyrus, Vossia* and other plants encloses the river on both shores and the fringe of vegetation offers shelter and food for a bewildering variety of epiphyton and almost all classes of animals with shoals of small fishes which find their livelihood there. A few meters outside this shore life currents of the midstream water cut off this community from up- and down-stream and also from free circulation of their metabolic products. The deeper benthos remains mostly stationary though some organisms are caught in the 'drift' and form the 'adventitious' element in plankton samples, together with detritus. In the Nile shore vegetation fades out towards the north, possibly because the river traverses arid climatic regions, created by past climatic changes. Not so in the Zaire and Amazon, where a variety of vegetational types occur in the floodplains for most of the river system, fulfilling the role of primary producers.

But even without this source of food, the bottom sediments in the Blue and White Niles produce a rich benthic fauna as seen every year by mass outbreaks of chironomids and in rocky regions of Simuliids.

Simuliids live in turbulent water and such 'rhithron' zones are interspersed in the 'potamon' courses of some of the rivers used as 'cases' in this essay, far away from any mountains. In the Zaire the cascades on the upper and lower part of the river create severe outbreaks of onchocerciasis. Similarly on the Nile such infestations occur in the cataracts of the

57

plains; fortunately these mass outbreaks consist mainly of *Simulium griseicolle*, which is not a vector.

Not enough is known about the bottom fauna of the three great rivers cited, but it must be rich judging from the 1,000 forms of chironomids recorded from the Amazon. There the richest habitat of invertebrate life exists in some of the 'floating meadows' which develop in particular parts of the river system. The mosaic of flowing and standing waters which is seen in all our three river systems represents fragments of the total potential of river life.

Fishes are mobile, yet their distribution in the Zaire and Amazon is subject to barriers and not to a zonation governed by temperatures mainly. In the Zaire extensive cascades separate to some degree the fish faunas of the Lualaba extension of the main river; similarly some 300 kilometers of turbulent waters prevent the entry of marine and euryhaline species into the river near the delta. Such restrictions on fish abundance in the Amazon are produced by limnological and especially chemical features of tributaries, some with pronounced acidity. In the Nile the Murchison falls separate the 'speciated' fish fauna of the lake Victoria from the poorer (in species) fauna of the downstream river; in lesser degree, the Blue Nile gorge separates the fishes of lake Tana from the river.

Development of plankton in rivers and its role as indicator

The most sensitive element to flow conditions is river plankton; I regard it as an acute indicator of the speed of water movement. Two main factors govern the appearance of the pelagic community, the velocity of currents with their differentiation and, intricately connected, and the time allowed for biological development.

J. F. Talling (J. F. T. & J. Rzóska 1967) has proposed some mathematical formulations for the development of plankton in moving water. I quote from this paper (pp. 655 and 657):

"The relative rate of population increase with distance downstream (i) will depend upon the interaction of the relative (specific) rates of net population increase with time (ii) and the flow velocity (iii). If the flow is homogeneous, with water moving at a single velocity without longitudinal mixing in eddies or storage-basins (i.e. 'plug' or 'piston' flow in continuous culture theory), the relationship of the three quantities is simply (i) = (ii)/(iii). Adopting symbols of N for population density, s for distance downstream, v for flow velocity, and t for time, and considering differential rates,

$$\text{(i) is } \frac{dN}{Nds}, \quad \text{(ii) is } \frac{dN}{Ndt}, \quad \text{(iii) is } v = \frac{ds}{dt}$$

and

$$\frac{dN}{Nds} = \frac{dN}{Ndt} \bigg/ \frac{ds}{dt}$$

58

The rates (i) and (ii) can also be expressed, in an integrated form, as ln units per unit distance and ln units per unit time respectively."

On p. 657 the mentioned paper examines the situation in 'mixed lake' conditions with through-flow:

"A quantitative difference between the plug-flow and mixed lake situations can be expressed as follows. Consider a river stretch of volume V, discharge v (dV/dt), an initial plankton concentration n_1, and a specific multiplication rate (ln units per unit time) of k. Under conditions of plug-flow, after a time period Δt the outflow concentration n_2 will be given by

$$\ln \frac{n_2}{n_1} = k\Delta t$$

whereas under mixed lake conditions, with a negligible plankton concentration in the inflow,

$$\ln \frac{n_2}{n_1} = \left(k - \frac{dV}{Vdt} \right) \Delta t$$

$$= \left(k - \frac{v}{V} \right) \Delta t$$

The equations above show that, for given values of v, V, and k, the specific rate of net population increase $[= \ln (n_2/n_1)/\Delta t]$ will be less under mixed-lake conditions than plug-flow conditions, the difference being one ln unit per mean retention time. Ecologically, this apparent penalty is offset by the potentially longer development time available under mixed lake conditions, the reduced dependence on upstream inocula, and the capacity for cumulative build-up of population density in one stretch of the river."

So much for the theory, the full discussion can be found in the quoted paper. If the speed of water flow governs the time allowed for the development of biological phenomena, then both a very low current velocity or the length of a river are important factors, in brief the 'age' of the water mass. Two examples from our Nile studies may illustrate the two points. The south-western tributary of the White Nile, the Ghazal river, carried in 1953/54 an almost 'pure' zooplankton in its lower course; this river is only 200 kilometers long but the *current* midstream, when measured, allowed for a flow of between 17 to 40 days. A graph (p. 18, Rzóska 1974) shows the dropping out of adventitious forms and the rise of true (Crustacean) plankton.

The second example comes from the Blue Nile and specifically the stretch of 357 kilometers from the Sennar dam to Khartoum. Samples taken at Khartoum and upstream in 1951–1955 during low water, that is mainly from March to May, showed the presence of a rich phyto- and zooplankton, the blue-green algae forming blooms, the crustaceans in full reproduction. The midstream current during the low season (outside the great flood from the end of June to the end of October) was on the average 0.1 m/sec and even lower; this would give a travel time of about 40 days for the 357 kilometers for this stretch. This then represents time allowed by *length* of river.

Two factors act against the permanence and full development of river plankton, a stronger current above 0.4 m/sec and the inadequate length of the river course. A number of observers have commented on the appearance of plankton organisms in flowing water: first diatoms and then rotifers appear. (Cilleuls, for French rivers, Margalef 1960 for Spanish running waters, van Oye 1926 for the Ruki river in the Zaire basin.) Hynes (1970) and Winner (in Whitton 1975) allude to this initial sequence. The recently best explored example is the Thames in England, where diatoms and rotifers appear but truly planktonic crustaceans are 'uncommon ... presumably cladocerans could not complete their life histories in plankton which reaches the sea in a week'. This is indeed an apt comment by the leader of the comprehensive study of the Thames at Reading (K. Mann in Oglesby et al. ed. 1973). R. Margalef (1960) has proposed a 'Synthetic approach to the Ecology of running waters'; he worked obviously on short rivers but he recognises that 'a river may need hundreds of kilometers to develop'. (I am indebted to Prof. Margalef for sending me his paper.)

Now it so happens that the Thames water is the main supply for a large part of metropolitan London and is pumped into reservoirs. There it clears of debris and a full development of phyto- and zooplankton occurs. Even more striking is the second case. Armitage (1977) studied 'drift' in two short streamlets joining the short river Tees in northern England. A dam was in action on one of the streams since 1970, and the author provided unintentionally an excellent proof for the development of plankton. He notes in the 'drift' at one time large numbers of *Daphnia hyalina, Bosmina coregoni, Chydorus sphaericus* and unnamed copepods. The dam was only 12 kilometers from the source of one of the streams. No data are given on time of storage in the dam basin nor on the fate of these truly planktonic organisms downstream. But this appearance, like in the Thames reservoirs, points to the importance of time, here divorced from currents, in other words 'retention' time.

Retention time of water is of paramount importance for biological phenomena. It occurs in all reservoirs on the Nile and no doubt over most of the world. The water flow loses its impetus, it clears, transparency increases and a biological sequence is set in motion, involving the production of phyto- and zooplankton, a specific bottom fauna, an increase of fishes, besides physical and chemical changes.

Not only dams create such sequence; natural obstacles such as rocky narrows stow up water; this must be the explanation of a dense *Melosira* bloom in the northern Nile near Dongola in winter 1957/8 with a rich cladoceran and copepod population. Other examples are the mouth bays of tributary rivers entering the Amazon; the stowing effect of one river on another entering is also known on the Nile. Some data on 'retention' time have been provided by hydrologists, e.g. in the Gebel Aulia dam on the White Nile has been calculated theoretically at about 40 days; the Aswan High Dam basin has a much longer retention time. In both the biological effects have been studied (Rzóska ed. 1976).

The question looms, how the supply of plankton in a river originates and how it can develop. The origin of river plankton was one of the early theoretical puzzles of hydrobiology. Inocula exist in quiet stretches, adjacent lagoons etc. Even during the enormous brown flood of the Blue Nile (June to October) some specimens of *Thermocyclops*, *Moina* and *Diaphanosoma* have been found, and in pools left by the falling river the speed of development was 2 or 3 generations in 10 days. The complete repopulation of the Blue Nile after the flood is a striking phenomenon. It occurs every year; A. Moghrabi (1977) thinks that resting stages are the main factor. The speed of growth and development of plankton crustaceans in moving water has not been studied. Only where this process could be followed in laboratory conditions or in small standing waters, some data have been obtained. H. Gauthier (1954) provided a great number of results on *Moina dubia* in Algerian waters, of which the following are quoted: some females produced their first batch of eggs two days after ecclosion; 50 specimens reared from ephippia produced in 4 days a population of 2,800 individuals in a temperature of 25°C. Similar results were obtained in rainpools around Khartoum for *Moina*, *Metacyclops* and conchostracans (Rzóska 1961). These examples, though not obtained in running waters, at least show the potential of organisms involved to respond to extreme conditions.

The study of productivity in rivers is therefore a difficult task and little known. The most intensive study is that on the Thames at Reading in England, where a team of botanists and zoologists tried to evaluate the production; this has been reported in a number of papers and by K. Mann in a concise survey (in Oglesby et al. ed. 1973). J. F. Talling has discussed the photosynthetic production in the Nile system (in Rzóska ed. 1976). For zooplankton this is much more difficult; even biomass assessments are made hazardous by the uneven densities floating by the observers point.

In general one can say that rivers, like all other water bodies are colonised by organisms intensively but selectively. Because of the different conditions present in the run of rivers, a mosaic of species develops in fragments with a gradual and fuller unfolding in long rivers. I call this the biological ontogeny.

The ontogeny of running waters

Three examples illustrate this concept, based on length, current and 'lifespan' of running waters and the biological response.
(a) Rheidol stream in Wales; length about 48 kilometers, mean velocity about 1 m/sec, total run calculated at about 13 hours; a rich benthos present.
(b) River Thames; length about 240 kilometers, average run of water to eastury 6–7 days; shore vegetation with fauna, benthos with mussels conspicuous; plankton of diatoms and rotifers mainly.
(c) River Volga, U.S.S.R., as in 1928 before dams; length 3,694 kilometers, time of flow in measured stretch of 2,747 kilometers was 50 days in flood, much longer in summer and autumn; full biological development of all components of aquatic life, including a rich plankton.

I regard these 3 examples as widely set markers for arranging running waters according to the time and space of existence, which forms the basis of their biological development. It would be useful to enlarge this series. I do not propose another classification scheme, but I think that studies on running waters should include the essential data on length and velocity of water movement, besides location, climate and geomorphology.

The concept of 'ecosystem' and its validity for running waters

This concept and the term of ecosystem was produced by the English botanist A. G. Tansley in 1935 and 1939. But before that, the interdependence of organisms inter se and with their place of life was recognised in terms of the 'biocoenosis' for oyster-beds by Möbius and by 'microcosm' for freshwater bodies by Forbes. This was by the end of the 19th century; later it was followed by a spate of terms, dealing with the grouping of organisms into 'communities', like 'socies, facies, phyto- and zoocoenose, chorocoenose, synusies' and others. These terms are evidence of grappling with the complexity of nature. Another set of concepts arose with the efforts to understand 'trophic levels' as basis for production processes and the 'flow of energy'.

In 1935 Tansley introduced the term 'ecological system' and in 1939 defined it as '. . . units of vegetation . . . and animals habitually associated with them and also all the physical and chemical components of the immediate environment or habitat, which form together a recognisable entity. Such a system may be called an ecosystem . . .'.

This term in its brief and appealing formulation spread throughout the world and was used for a variety of situations. H. T. Odum defined in 1963 the ecosystem again as 'basic functional unit of nature, which includes both organisms and their non-living environment, each interacting with others and influencing each other's properties and both necessary for the maintenance and developments of the system'. This quotation is taken from a book on the application of the concept for the management of natural resources (van Dyne ed. 1969). There further definitions are added, 'integrated complex' and including the '. . . transformation, circulation and the flow of energy through and within living organisms'. On p. 309: '. . . a river . . . catchment . . . is a specific segment of the earth's surface, set off from adjacent segments of more or less defined boundary and occupied at any given time by a particular grouping of plants and animals'. Am I wrong to discern here some difficulty of allocating natural phenomena into human concepts? Boughey (1971) though accepting the ecosystem concept, declares that it is arbitrary; it can feature the whole 'ecosphere' of the earth or small situations like water containers of a Bromeliad plant or the assembly of organisms 'around a mouse pellet'. More of such pronouncements could be assembled. It is obvious that the term 'ecosystem', originally coined for a vegetation patch, easily recognisable, and its animal etc. components has deviated and is used indiscriminately almost like a 'slogan'. A very critical re-examination of its use should be useful.

62

Rivers are not 'ecosystems'

During a symposium on 'River ecology and Man' (Oglesby et al. ed. 1973) one of the session chairmen remarked, winding up, that rivers should be regarded 'above all' as ecosystems. This in my opinion is meaningless. In previous pages I tried to show, that in flowing water recognisable biological formations are not integrated in a flux of dependencies; e.g. the papyrus fringe of the Upper Nile does not share its food web and its metabolic products with other quite different sectors of the river. The documentary photographs in the Nile section of this essay show a linear arrangement of riverain landscapes and habitats, following each other like 'beads on a string' using Gessner's apt expression. I do not go as far as Gessner to say that the only unity in a river is its name. The river in flow binds the landscapes of the valley together and the water flowing past Cairo comes from sources thousands of kilometers upstream. It mixes and carries away local influences without allowing a circulation of matter or 'energy flow'. The same happens probably in other long rivers like the Zaire and the Amazon.

This has already been recognised by some critical observers. Schwörbel in a review of current literature mainly on streams (1969) underlines the 'continuous downflow of nutrients' in contrast to stationary circulation in lakes. Hynes (1970, p. 410) remarks that the self-contained complex of trophic levels and their circulation 'clearly does not apply to the ecosystems in running waters, where everything released into solution ... has little opportunity of being recycled on the spot'. Sioli, with his rich experiences on the great Amazon, says '... rivers are through-flow systems .... Rivers are nothing else than functional parts of higher units ... expressions of their terrestrial environment ...'. A panel of limnologists on project 5 of the MAB (Man and the Biosphere) programme concluded that a 'river system ... may be a higher level of organisation than the ecosystem' (Unesco, Paris 1972). I asked also for the private opinions of three scientists on the ecosystem concept for rivers. Prof. Hynes wrote that the term is not tenable for long rivers especially those passing through different climatic zones like the Nile; Prof. Sioli extended his doubts on the applicability to all rivers; he regards 'all these terms as more or less arti-ficial'. Prof. Illies replied that with the river ecosystem everything depends upon the definition of 'ecosystem'. Even long rivers form along their course 'eine einheitliche Grundsituation' and he does not see difficulties in applying the term. Indeed, there is the binding effect of the water medium, its floating suspension of plankton and the mobile fishes. But in no way does a river, especially a long one, conform to any definition of ecosystem now current. No term exists as yet to express the nature of rivers adequately.

Desiderata for future work on running waters

1. Any study on running waters should above all include information on

63

length, current velocity, in brief the duration of the flow.

2. Biologists should be acquainted with the essentials of hydraulics and hydrology.

3. Though it is a prerequisite in science that results of previous studies should be known, yet as shown in this essay such results have been ignored by some recent opinions.

4. Books based on symposia consist often of disjointed and contradictory contributions; working groups should clarify such discrepancies. Otherwise embarassing incidents may occur of lack of percolation even across the pages of one book, e.g. in Oglesby et al. ed. (1973).

5. It is also evident that international diffusion and exchange of results suffers from insularism of big countries and ignorance of languages other than one's own.

6. Finally, the heavy terminological 'armour' of present ecological science often prevents an unbiassed look at nature. I cannot refrain from quoting a delightful story recently heard; a professor of a big institution in a big country urged his disciples not to go into the field before they had a 'model'. An excellent Italian proverb says 'If it is not true, it is well invented'.

I end these reflections with a request to the reader to contemplate the saying by Leonardo da Vinci, which I have chosen as Motto of this essay.

References

For the Nile only few references are given, which are necessary and support the opinions expressed in this booklet. The other 500 sources, on which the book on 'The Nile Biology of an ancient river' (1976) was based are not quoted here.

Ambühl, H. 1959. Die Strömung als ökologisher Faktor. *Schweiz. Zeitschr. f. Hydrologie* 21.

Armitage, P. D. 1977. Invertebrate drift in the regulated river Tees, and an unregulated tributary, Maize Brook, below the Cow Green dam. *Freshwater Biology* vol. 7, 2.

Behning, A. 1928. *Das Leben der Wolga. Zugleich eine Einführung in das Studium der Flussbiologie.* Die Binnengewässer Band V. Schweizerbart, Stuttgart.

— 1929. Das Plankton der Wolga. Verhdlg. S.I.L. Congress 4. Rome

Berg, A. 1959. Analyse des conditions impropres au développement se la jacinthe d'eau Eichhornia crassipes (Mart) Solms dans certaines rivières de la cuvette congolaise. Bull. Agricole du Congo Belge et du Ruanda-Urundi vol. 1 Bruxelles.

— 1961. Rôle écologique des eaux de la Cuvette congolaise sur la croissance de la jacinthe d'eau *Eichhornia crassipes*. Acad. Roy. des Sciences d'Outre-Mer, Tome XII, fasc. 3 Bruxelles.

Bont, A. F. De 1969. The status of limnological knowledge of the Congo basin in: Report of the Regional Meeting of hydrobiologists in tropical Africa (organised by the Intern. Biol. Programme), Uganda 20–28 May 1968. Publ. by Unesco Regional Centre, Nairobi, Kenya.

— & H. Maes 1956. Conductivité électrique de quelques rivières Katangaises *Hydrobiologia* 8.

Boughey, A. S. 1971. *Fundamental Ecology.* Intext Educational Publishers, Saranto-San Francisco-Toronto-London.

Butcher, A. D. 1940. Studies in the ecology of rivers–4. Observations on the growth and distribution of the sessile algae in the river Hull, Yorkshire. *J. Ecology* 28, 210–223.

Butzer, K. W. 1966 Climatic changes in the arid zones of Africa during early to mid-Holocene times. Roy. Meteor. Soc.: World Climate from 8000 to 0 B.C. Proc.

Carpenter, K. 1928. *Life in inland waters.* Sedgewick & Jackson, London.

Cilleuls, J. des. 1928 Revue générale des études sur le plankton des grands fleuves et rivières. *Intern. Rev. Hydrobiol. Hydrographie* 20.

Coker, R. E. 1954. *Streams, lakes and ponds.* Univ. of North Carolina Press.

Eddy, S. 1932. The plankton of the Sangamon river in the summer of 1929. State Illinois Div. Nat. Hist. Survey 19.

Einsele, W. 1957. Flussbiologie, Kraftwerke und Fischerei. *Schrift. Österr. Fischereiverbands* 1.

Emery, W. B. 1967. *Archaic Egypt.* Pelican Book, London.

Fittkau, E. J. 1970. Role of caymans in the nutrient regime of mouth-lakes of Amazon affluents (A hypothesis). *Biotropica* 2.

— 1974. Zur ökologischen Gliederung Amazoniens. I Die erdgeschichtliche Entwicklung Amazoniens. *Amazoniana* 5.

—, & J. Illies, H. Klinge, G. H. Schwabe, H. Sioli 1968, 1969. *Biogeography and Ecology in South America.* Monographiae Biologicae vol 18 & 19; Junk publ. The Hague.

Forbes, S. A. & R. B. Richardson 1913. Studies in the biology of the Upper Illinois river. *Bull. Illinois Lab. Nat. Hist.* 9.

—, —, 1919. Recent changes in Illinois river biology *Bull. Illinois Lab. Nat. Hist.* 13.

Forbes, S. A. 1927–28. The biological survey of a river system, its objects, methods and results. *Bull. Illinois State Nat. Hist. Survey Div.* 17.

Galtsoff, P. S. 1924. Limnological observations in the Upper Mississippi. *Bull. U.S. Bureau Fisheries* 39.

Gauthier, H. 1954. *Essai sur la variabilité, l'écologie, le determinisme du sexe et la réproduction de quelques Moina recoltées en Afrique et à Madagascar.* Algiers, private print.

Gessner, P. 1955. *Hydrobotanik. Die physiologischen Grundlagen der Pflanzenverbreitung in Wasser. I Energiehaushalt.* VEB Deutcher Verlag der Wissenschaften. Berlin.

— 1964. The limnology of tropical rivers. Verhdlg. S.I.L. 15.

Hammerton, D. 1972. The Nile River—a case history in River Ecology and Man, Oglesby *et al.* edit.

Hurst, H. E. 1952. *The Nile, a general account of the river and the utilization of its waters.* Sec. ed. 1957. Constabte, London.

Hynes, H. B. N. 1970. *The ecology of running water.* University of Toronto Press.

Illies, J. & L. Botosaneanu 1963. Problèmes et méthodes de la classification des eaux courantes, considerées surtout du point de vue faunistique. *S.I.L. Mittlg.* 12.

Kassas, M. 1972. Ecological consequences of water development projects. The Environmental Future, ed. N. Polunin, MacMillan Press, London.

Kofoid, C. A. 1897–1910. Plankton Studies. Part 1–5 under several subtitles. *Bull. Illinois State Lab. Nat. Hist.* vol. 5–8.

Liepolt, H. (ed) 1967. *Die Limnologie der Donau.* Stuttgart.

Lowe-McConnell, R. H. *Fish Communities in Tropical Freshwaters.* Longman, London & New York.

Lauterborn, R. 1916–1918. Die geographische und biologische Gliederung des Rheinstromes. Sitz-ber. d. Heidelberg Akad. Wissenschaften, Math.-Nat. Klasse Abtlg. B., I Teil-1916, II 1917, III 1918.

Lyons, H. G. 1909. Longitudinal Section of the Nile. *Geogr. J.* 34:36–51.

M.A.B. 1972. Man and the biosphere. Expert panel on Project 5: Ecological effects of human activities on the value and resources of lakes . . . rivers . . . Final Report. Unesco Paris.

Magis, N. 1964. Étude limnologique des lacs artificielles de la Lufira et du Lualaba (Haut Katanga). II Étude chimique des eaux de la retnue de la Lufira (lac de Mwadingusha). *Bull. Soc. Roy. Sciences Liège* 45.

Mann, K. N. 1973. Case story: The river Thames. In: Oglesby et al.

Margalef, R. 1960. Ideas for a synthetic approach to the ecology of running waters. *Intern. Rev. ges. Hydrobiologie* 45.

Marlier, G. 1951. Recherches hydrobiologiques dans les rivières du Congo oriental. Composition des eaux. La conductibilité électrique. *Hydrobiologia* 3.

— 1954. Recherches hydrobiologiques dans les rivières du Congo oriental. II Étude écologique. *Hydrobiologia* 6.

— 1958. Recherches hydrobiologiques au lac Tumba (Congo Belge, prov. de l'Equateur). *Hydrobiologia* 10.

— 1973. Limnology of the Congo and Amazon rivers. In: Meggers et al.: Tropical forest ecosystems in Africa and South America. Smithsonian Inst. Princeton.

Moghrabi, Asim el 1977. A study of diapause of zooplankton in a tropical river – the Blue Nile. *Freshwater Biology* 7, 3.

Nile water and lake dam projects 1976. Preprints of papers presented at the symposium held at the National Research Centre, Dokki, Cairo, 1–4. March 1976.

Oglesby, R. T & C. A. Carlson, J. A. McCann (edit.) 1972. *River ecology and Man.* Environmental Sciences, Academic Press, New York–London.

Oye, van P. 1926. Le potamoplankton du Ruki au Congo Belge et des pays chauds en général. *Intern. Rev. Hydrob. & Hydr.* 16.

Poll, M. 1957–1963. ⎫
Poll, M. et J. P. Gosse 1963. ⎬ Quoted after R. H. McConnell 1975.

Robert, M. 1942. Le Congo Physique. Deuxième édition, Bruxelles, Librairie des Sciences.

Rzóska, J. 1974. The Upper Nile swamps, a tropical wetland study. *Freshwat. Biology*, vol. 4: 1–30.

— edit. 1976. *The Nile Biology of an Ancient River.* Monographiae Biologicae 29, Junk Publishers, The Hague.

— 1976. A controversy reviewed, Aswan High Dam. *Nature,* 261, no. 5560.

Schwörbel, J. 1969. Ökologie der Süsswassertiere. *Fortschritte der Zoologie* 20.

Sioli, H. 1956. Über Natur und Mensch im Brasilianischen Amazonasgebiet. Erdkunde-Archiv f. wissenschaftliche Geographie, Bonn 10, 2.

— 1966. Soils in the estuary of the Amazon. Humid Tropic Research, Proc. Dacca Symposium, UNESCO.

— 1966. General features of the delta of the Amazon. *Ibidem.*

— 1967. Studies in Amazonian waters. Atas de Simposio sobre a Biota Amazonica vol. 3, Limnologia.

— 1973. Recent human activities in the Brasilian Amazon region and their ecological effects. In: Meggers et al. edit. Tropical Forest Ecosystems in Africa and South America. Smithsonian Inst., Princeton 8.

— 1975. Tropical River: The Amazon (case story). In: Whitton et al. River Ecology, Blackwell Oxford.

— 1975. Amazon tributaries and drainage basin. In: A. D. Hasler edit. Coupling of land and water systems Springer Publ.

— 1977. Amazonien, Der Welt grösster Wald in Gefahr. Umschau in Wissenschaft und Technik 77, 5.

Sternberg, H. O'Reilly 1960. Radiocarbon dating as applied to a problem of Amazonian morphology. Comptes Rendus 18 Congrès Int. de Géographie.

Symoens, J. J. 1968. La minéralisation des eaux naturelles, conçernant les eaux courantes du Katanga et aussi du Congo. Édition Cercle Hydrobiologique de Bruxelles.

Talling, J. F. & J. Rzóska 1967. The development of plankton in relation to hydrological regime in the Blue Nile. *J. Ecol.* 55:657–662.

Talling, J. F. 1976. Phytoplankton: Composition, development and productivity. In: J. Rzóska edit. The Nile Biology of an Ancient River. Junk Publ. The Hague.

Tansley, A. G. 1935. The use and misuse of vegetational terms and concepts. *J. Ecol.* 16.

— 1939. *The British Islands and their vegetation.* Cambridge University Press.

Van Dyne, G. M. edit. 1969. *The ecosystem concept in natural resources management.* Academic Press New York–London.

Wickens, G. 1975. Changes in the climate and vegetation of the Sudan since 20,000 BP. Proc. 8th Plen-Session AETFAT, Genève 1974, Boissiera.

Willcocks, W. 1904. The Nile in 1904. National Printing Dept. Cairo.

Whitton, B. A. edit. *River Ecology.* Studies in Ecology vol. 2, Blackwell Oxford.